WHERE HAVE ALL THE BIRDS GONE?

Nature in Crisis

Rebecca E. Hirsch

TWENTY-FIRST CENTURY BOOKS / MINNEAPOLIS

To Rick, for his never-ending support

Twenty-First Century Books™
An imprint of Lerner Publishing Group, Inc.
241 First Avenue North
Minneapolis, MN 55401 USA

For reading levels and more information, look up this title at www.lernerbooks.com.
Illustrations on pp. 6, 15, 29, and 74 by Laura K. Westlund.

Main body text set in ITC Caslon 224 Std.
Typeface provided by Adobe Systems.

Library of Congress Cataloging-in-Publication Data

Names: Hirsch, Rebecca E., author.
Title: Where have all the birds gone? : nature in crisis / Rebecca E. Hirsch.
Description: Minneapolis : Twenty-First Century Books, [2022] | Includes
    bibliographical references and index. | Audience: Ages 13–18 | Audience: Grades
    7–9 | Summary: "In the face of rapidly declining bird populations, read about the
    vast impacts birds have on ecosystems, food systems, and our mental health and
    what we can do to protect them"— Provided by publisher.
Identifiers: LCCN 2021039521 (print) | LCCN 2021039522 (ebook) |
    ISBN 9781728431772 (library binding) | ISBN 9781728445441 (ebook)
Subjects: LCSH: Birds—Conservation—Juvenile literature. | Bird populations—
    Juvenile literature.
Classification: LCC QL676.5 .H57 2022 (print) | LCC QL676.5 (ebook) | DDC
    639.9/78—dc23

LC record available at https://lccn.loc.gov/2021039521
LC ebook record available at https://lccn.loc.gov/2021039522

Manufactured in the United States of America
1-49564-49543-10/6/2021

Contents

THREE BILLION TO ZERO

I have stood for hours admiring the movements of these birds. I have seen them fly in unbroken lines from the horizon, one line succeeding another from morning until night.

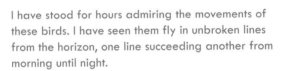

—Potawatomi activist and writer Simon Pokagon

In May 1850, Simon Pokagon stepped out of his shelter. The twenty-year-old Potawatomi tribal leader was camped out near the headwaters of the Manistee River in northern Michigan. There, a loud, strange rumbling sound greeted him. It sounded like an army of horses with sleigh bells advancing through the woods. Pokagon listened carefully. No, he decided, the noise wasn't the beating of horse hooves and the ringing of sleigh bells. It must be distant thunder. Yet the sky was clear.

Nearer and nearer came the mysterious sound. At last, he saw the source of the rumbling—a flying mass of millions of pigeons.

He stood as still as a statue, with birds flying all around him. "They passed like a cloud through the branches of the big trees, through the underbrush and over the ground," he later wrote. They were passenger pigeons (*Ectopistes migratorius*), pretty birds a little bigger than mourning doves. The males were slate blue on top with coppery undersides. The females were brownish. The birds flapped their long, tapered wings and landed all around—on the branches, on the ground, on his head, on his shoulders. The fluttering of their wings and the ringing of their chatter were deafening.

American artist John James Audubon made this painting of male and female passenger pigeons in the early 1800s.

Pokagon had watched passenger pigeons all his life. In the early nineteenth century, when he was a boy, they were likely the most numerous birds on the planet. Year after year, roughly three billion passenger pigeons migrated north and south through the eastern and midwestern United States and Canada. They flew as far north as Ontario, Quebec, and Nova Scotia and as far south as Texas, Louisiana, Alabama, Georgia, and Florida.

The size of the flocks was staggering. They were so immense that they darkened the skies for days as they passed overhead. One flock, estimated at one billion birds, stretched 300 miles (483 km)

Lake Superior

CANADA

MICHIGAN (Upper Peninsula)

Lake Huron

Manistee River

WISCONSIN

Lake Michigan

⅄ = 20 million pigeons

N

MICHIGAN

Miles
0   50   100

0   50   100   150
Kilometers

Billion-Bird Migration

The flock of passenger pigeons witnessed in 1850, if viewed from above, would have stretched approximately this far across northern Michigan.

long and took fourteen hours, from sunup to sundown, to pass overhead. Many years later, Pokagon described watching the birds flow "like some great river" across the sky.

Before the nineteenth century, passenger pigeons lived a secure existence in North America. They migrated through the eastern forests, searching for the acorns and other nuts that nourish them. Each spring they headed to the Great Lakes region to breed. The birds that Pokagon witnessed on the Manistee River were part of an enormous flock descending into the forest to mate and nest.

In 1895, as Pokagon neared the end of his life, he recalled, "I have stood by the grandest waterfall of America . . . yet never have my astonishment, wonder, and admiration been so stirred as when I have witnessed these birds drop from their course like meteors from heaven." But by the time he wrote those words, passenger pigeon flocks numbered in the dozens rather than the millions or billions. The birds were almost gone.

Where did they all go?

THE LAST OF THE PASSENGER PIGEONS

In the nineteenth century, passenger pigeons collided with two deadly forces: overhunting and destruction of their forests. The large flocks often damaged food crops, so farmers retaliated by shooting the birds. Some hunters killed the pigeons merely for sport. Others shipped them across the country to be sold as food. And hunting passenger pigeons was easy. The flocks were so thick and so vast that a hunter could easily shoot a thousand birds in one outing.

By 1850, the year Pokagon was camping on the banks of the Manistee River, hunters were killing the birds faster than they were reproducing. As their numbers dwindled, some states passed laws to limit or restrict the hunting of passenger pigeons, but people widely ignored the laws. Meanwhile, loggers were clearing woods to make way for cities and farms. The trees they cut down were used as building material and burned for fuel. As large tracts of forest disappeared from the East and Midwest, passenger pigeons lost their habitat.

The population collapsed in a downward spiral and never recovered. The birds' destruction alarmed some people. "The wild pigeon, formerly in flocks of millions, has entirely disappeared from the face of the earth," said US representative John F. Lacey of Iowa. "We have given an awful exhibition of slaughter and destruction, which may serve as a warning to all mankind." Lacey introduced

This illustration from the 1870s shows American hunters shooting passenger pigeons.

the first national wildlife protection law, which Congress enacted in 1900. The Lacey Act made it a federal crime to sell illegally hunted game across state lines. The law came too late to save passenger pigeons. In 1902 a hunter in Indiana shot a passenger pigeon in the countryside. After that, no one saw any more passenger pigeons in the wild, although some lived in zoos. In 1909 the American Ornithologists' Union, an association of scientists who studied birds, offered a $3,000 reward to anyone who could locate nesting passenger pigeons. The search lasted for three years, but no nests or birds were found. The very last passenger pigeon, a captive bird called Martha (named for US first lady Martha Washington), died in her cage on September 1, 1914, at a zoo in Cincinnati, Ohio. In less than a single human lifetime, the population of passenger pigeons had gone from three billion to zero.

THE SIXTH EXTINCTION

In one sense, the story of the passenger pigeon isn't unique. The history of life on Earth is riddled with extinctions. Extinctions often occur when the environment changes. If an area becomes drier, for instance, water-loving plants and animals might not be able to survive there. Their species might die out.

Sometimes vast numbers of species die out suddenly, all around the same time. Scientists call this event a mass extinction. Mass extinctions usually follow a dramatic, planetwide change. For instance, about sixty-five million years ago, an asteroid slammed into a shallow sea near the Yucatán Peninsula of Mexico. The impact created a giant crater, churned up tidal waves, and kicked trillions of tons of red-hot dust into the air. Winds spread the smoldering dust around the globe, and forests worldwide caught on fire. With so much smoke, dust, and debris in the air, the skies grew dark for months. Plants need sunlight to make food, but under the darkened skies, most of the world's plants could not survive. Without plants, many animals didn't have enough to eat, so animal species died out too. Altogether, the event caused the extinction of about 75 percent of all species on Earth, including nearly all the dinosaurs.

Called the Cretaceous-Tertiary extinction, this was the fifth mass extinction in Earth's history. Scientists say that in modern times, we're in the middle of a sixth extinction. The cause this time? *Homo sapiens*, the scientific name for humans.

Humans have come to dominate life on Earth and have dramatically changed the planet. We have driven many species to extinction. Overhunting caused the extinction of some animals. Other extinctions occurred when people cleared land to build farms and cities. This destroyed the natural homes of many plants and animals, leading some species to die out. People have also released dangerous pollutants into the air and water, killing many plants and animals and leading to even more extinctions.

One of the most harmful forms of pollution is carbon dioxide, a type of gas. Carbon dioxide enters the air whenever we burn fossil fuels—coal, petroleum, and natural gas. For instance, petroleum-powered cars, gas-burning furnaces, and coal-fueled power plants all emit carbon dioxide. In the atmosphere, carbon dioxide acts like the glass roof and walls of a greenhouse, trapping heat and causing temperatures to rise. The higher temperatures are melting ice in polar regions and changing Earth's climate. Many dry areas are getting drier. Many wet areas are getting wetter. Storms, floods, droughts, and other weather extremes are becoming more severe. Climate change is happening quickly, and birds and many other

SURVIVING THE FIFTH EXTINCTION

During the Cretaceous-Tertiary extinction, most dinosaur species died out. But some dinosaurs—the ancestors of modern birds—survived the extinction. No one knows exactly how they survived on the dark, deforested planet after most other plants and animals had perished. They probably eked out a living by eating seeds they found on the ground.

Ancient birds were not the only survivors of the catastrophe. A few small mammals and some marine animals also survived. Eventually, the skies cleared and life on Earth recovered. Over millions of generations, plants and animals evolved, changing and diversifying to become millions of new species. The ancient birds became eleven thousand species of modern birds. The small mammals that had survived the extinction also changed. They evolved to become the many mammal species that populate modern Earth. These mammals include humans.

When people clear land to build roads and structures, they destroy bird habitat.

animals are struggling to adapt. Because of climate change, many species are at risk of dying out.

According to the International Union for Conservation of Nature, a Swiss organization that works to protect wildlife, at least 5 percent of plants and at least 14 percent of vertebrates (animals with backbones) are threatened with extinction. Among birds, 14 percent of species are at risk of extinction. The passenger pigeon is gone forever, and many other birds are in danger. Can we save the birds that still live with us?

Three Billion to Zero

DISAPPEARING BIRDS

In pushing other species to extinction, humanity is busy sawing off the limb on which it is perched.

—American biologist Paul R. Ehrlich

Birds are everywhere. You see them hopping on city sidewalks. You hear them singing on farms and in forests. You spot them circling over lakes and marshes. If you were to go for a walk, you'd likely encounter a bird. Maybe you'd spot a robin yanking a worm out of the ground. Or hear a songbird singing. Or spy a solitary hawk hunting from a high branch. Birds are everywhere, so it's easy to assume they'll always be there.

It may sound unlikely, even impossible, but our birds are in trouble. Scientists have gathered clear and alarming evidence that many species of birds are disappearing in North America and around the world.

Barnacle geese fly over wetlands.

 ## WARNING SIGNS

In the twentieth century, ornithologists saw warning signs that North American birds might be in trouble. Many longtime bird-watchers reported seeing fewer birds than decades before. And from long-term bird surveys, or bird counts, scientists knew some species were shrinking in number. But a few scientists pointed out that other bird species were *growing* in number.

What was really happening? Was the total number of North American birds increasing, decreasing, or staying steady? How were our birds doing as a whole?

In 2007 a team of researchers decided to find out. They came from universities, government agencies, and nonprofit groups in the United States and Canada, and the approach they took was exhaustive. First, they combed through bird surveys going back to 1970. During these surveys, ornithologists and amateur bird-watchers had counted bird populations year after year in thousands of locations across the United States and Canada.

FLOATING ANGELS

Radar was first used during World War II (1939–1945). Early in the war, British troops used radar to look for enemy aircraft in flight. They noticed mysterious shadows floating across their radar screens. They knew the shadows weren't created by aircraft or rain and snow. The source of the shadows was a mystery. Radar operators nicknamed them angels.

The mystery of the angels was finally solved in 1958 by a New Orleans, Louisiana, high school student named Sidney Gauthreaux. His hometown lay in the path of a busy flyway (a kind of superhighway for birds). As a child, he would lie awake, listening to the calls of birds passing in the night. In the 1950s, weather agencies installed radar stations along the coast of the Gulf of Mexico, near Gauthreaux's home. He wondered if the weather radar signals would bounce off the bodies of the birds he heard at night. In 1958 station operators allowed him to look at some of their weather radar images. There he saw the "angels" that had puzzled British radar operators. He realized that those mysterious shadows were really clouds of migrating birds.

Gauthreaux went on to become a scientist and continued using radar to study birds. As a graduate student at Louisiana State University in the 1960s, he used radar images to document that birds were flying in large numbers across the Gulf of Mexico, something that was unknown at the time. In the late 1980s, he made another startling discovery: bird migrations across the Gulf had shrunk by half since the 1960s.

Scientists and others continue to use radar to study and conserve birds. Radar images reveal where birds stop and rest along their migratory routes. This information is used to protect key habitat, ensuring that birds can find food and shelter as they travel between their summer and winter

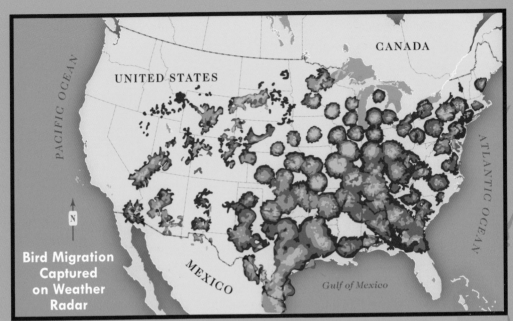

Bird Migration Captured on Weather Radar

Source: US Geological Survey (USGS), https://www.usgs.gov/media/images/nexradcompositejpg

homes. At airports, radar images can show where large numbers of birds might cross the flight path of an airplane. Air traffic controllers can then divert the plane to prevent a collision. Bird-watchers can check online radar images to track bird migration in real time. They can see when birds will be nearby and make plans to watch them through binoculars.

Savvy radar watchers know the clues that distinguish birds on radar screens from rain and snow. Birds show up as blobs that bloom in size, radiating outward. Each blob is a large group of birds taking off and spreading out as they fly. The blobs tend to show up on radar screens about a half hour after sunset, when migrating birds usually take off. They disappear just before sunrise, when birds stop to rest during the day.

These surveys gave the team data on 529 North American bird species. The data revealed whether each species was rising, falling, or holding steady in number.

Then they examined images from radar stations. Weather forecasters use radar to track rain and snow. At weather stations, radar systems send out radio wave pulses. The pulses bounce off raindrops and snowflakes and return to the station. Radar equipment then turns the returning signals into images, showing the location of wet weather over a large area. But radio waves bounce off raindrops and birds alike, and large masses of migrating birds often show up on weather radar. So the research team studied radar data from 143 weather stations across the United States. They looked at birds that showed up on weather radar during spring migrations between 2007 and 2017. With this information, the team could track how the size of the North American spring bird migration had changed over time.

In October 2019, the team published its report in the journal *Science*. The conclusion was startling. Since 1970 nearly three billion birds—nearly 30 percent of all the birds in the United States and Canada—had vanished.

 ## A FULL-BLOWN CRISIS

Three billion birds, gone in just fifty years. The researchers were shocked. "We were stunned by the result—it's just staggering," said lead author Ken Rosenberg of the Cornell Lab of Ornithology at Cornell University in New York.

David Yarnold, then president and chief executive of the National Audubon Society, a national nonprofit organization dedicated to bird conservation, called the findings "a full-blown crisis."

Birds had disappeared from almost every habitat: Forests had lost 1 billion birds, grasslands had lost 720 million birds, Arctic tundra had lost 80 million birds, and coasts had lost 6 million birds. Aerial insectivores—insect-eating birds such as swallows and swifts—were down by 160 million, or 32 percent. Ordinary backyard birds, such as robins and sparrows, had been hit with big losses. So had familiar visitors to bird feeders, including blue jays and juncos.

THEY DON'T KNOW WHAT THEY'RE MISSING

What does it look like when three billion birds disappear? The truth is, it's quiet and undramatic. The loss has taken place gradually, over fifty years, so it can be hard to detect. "Birds are not dropping out of the sky," said Cagan Sekercioglu, an ornithologist at the University of Utah in Salt Lake City, so the loss can be hard to detect.

"There are still a lot of birds out there," Rosenberg explained. "If you have a lot of birds coming to your feeder and they're reduced by thirty percent, you might not see that."

Over many decades, each new human generation grows up seeing fewer birds than their parents saw. Each generation grows up thinking the amount of birds they see is normal. Scientists have a name for this phenomenon: shrinking baseline syndrome. As Sekercioglu explains, "When you are young," the number of birds you see is "your baseline [normal]. The problem is, the next generation, their baseline is lower. But they don't know what they're missing."

THE BLUEBIRD OF HAPPINESS

Why do shrinking bird populations matter? If billions of birds are still out there, why should we care if some are gone? We should care because all other living things need birds.

Birds are nature's essential workers. When they disappear, so do the services they provide in nature. All plants and animals belong to complicated webs called ecosystems. Birds perform services that ecosystems need and that we humans also depend on. For instance, hummingbirds are pollinators. When they visit flowers to drink nectar, they come away dusted with pollen and carry this pollen to the next flower they visit. This transfer of pollen allows plants to make seeds and reproduce. Swifts are exterminators, eating insects that can destroy food crops. Hawks prey on rodents, which can contaminate our food supplies, damage property, and spread diseases if they aren't kept in check. Vultures are cleanup crews, scavenging waste such as roadkill that would otherwise pile up outdoors and possibly spread disease.

A hawk soars on a column of warm air.

Birds are also bridges. They connect us to the natural world. They delight us with their bright colors, their beautiful songs, and their powers of flight. They connect us to the passing of time and the change of the seasons, from the first robin of spring to honking geese flying south for the winter.

When birds disappear, our connection to nature—already frayed by life in the modern era—weakens further. But when our connection to nature is strong, it helps keep us happy. Scientific studies have shown that exposure to nature, and especially birds, is good for your mental and emotional well-being. One study from Germany found that people in Europe who lived near more bird species were happier. A study in the United States revealed that hikers who listened to birdsong while walking along a trail experienced greater joy than other hikers. When birds disappear, we lose their uplifting influence on our lives—the joy of hearing their chorus, the delight of seeing them soar.

THREE BILLION CANARIES

In the 1800s and early 1900s, coal miners in North America and Europe carried caged canaries with them into the mines. Canaries are very sensitive to the effects of carbon monoxide, an invisible, odorless, and deadly gas that can occur naturally in mines. If a canary swayed and fell off its perch, it was a clear signal to the miners that poisonous gas was present. The miners quickly evacuated.

Like the dead canaries, disappearing birds are giving us a clear warning. Shrinking bird populations are an early signal that something is seriously wrong with our environment. It appears to be growing unsafe—not just for birds but for other living things.

Birds are disappearing not only in North America but around the world. A 2018 study by the group BirdLife International found that of all the world's bird species, at least 40 percent are shrinking in number. Just as in North America, the losses cut across many kinds of birds and many types of habitats. With these losses, warning lights are flashing and sirens are wailing for ornithologists and bird lovers. What is happening to our beloved birds?

BIRD COMEBACK

Don't count birds out. The good news is that bird populations can return to vibrant health if given a helping hand. Want proof? Just look in the *Science* report describing the loss of nearly three billion North American birds. That report also identified two bird groups that have made stunning turnarounds.

The report showed that ducks, geese, swans, and other wetland birds are on the upswing. Wetlands, where the ground is covered or saturated with water, include wooded swamps, marshes, and bogs; floodplains along rivers and streams; and the fringes of lakes and ponds. Early in US history, people began to drain wetlands to build farms and towns. By 1984 more than half of all wetlands in the United States had been cleared of water. Others had been polluted by chemicals from industry and agriculture. With their habitats destroyed, populations of waterfowl dropped to historic lows.

Protected wetlands like this one in Texas are home to ducks and many other birds.

Finally, conservationists called for wetland protection. Hunters also wanted healthy wetlands, where they could kill waterfowl responsibly and within legal limits. The two groups teamed up and pressured politicians to pass laws protecting wetlands. Some laws make it illegal to drain wetlands to build farms, roads, or other structures. Others designated funds to clean up wetlands that had been polluted and to restore wetland plants. With such habitat protections in place, dwindling waterfowl populations have bounced back.

Birds of prey are thriving too. Some birds of prey—including bald eagles, peregrine falcons, and ospreys—have returned to the skies after nearly going extinct in the twentieth century. This is because people worked to pass laws and set up programs to support the birds. For instance, some groups bred these birds in captivity to increase their numbers and then released the young birds into the wild. Laws that protect birds of prey and their habitats include the Migratory Bird Treaty Act, the Bald and Golden Eagle Protection Act, the Clean Water Act, and the Endangered Species Act.

These success stories drive home a crucial point: conservation works. "When people band together and take action, it is possible to reverse population declines and bring species back from the brink," said Nicole Michel of the National Audubon Society. "Birds are down, but they're not out. When you give them half a chance, they can recover." Although about 30 percent of our birds have been lost since 1970, 70 percent remain. And that's enough to jump-start a recovery if we act on conservation now.

"The story is not over," said Michael Parr, president of American Bird Conservancy, a Virginia-based nonprofit dedicated to conserving wild birds and their habitats. "There are so many ways to help save birds."

A CLEAR DANGER

Look closely at nature. Every species is a masterpiece, exquisitely adapted to the particular environment in which it has survived. Who are we to destroy or even diminish biodiversity?

—American biologist E. O. Wilson

Professor Daniel Klem Jr. slides open a drawer full of dead birds. We are standing inside the Acopian Center for Ornithology at Muhlenberg College in Allentown, Pennsylvania, where Klem is a professor. He picks up a lifeless bird of prey, a Cooper's hawk, and holds it under the fluorescent lights. He explains that the hawk was chasing two birds when it smashed into a window. Then he points to a merlin, a smaller bird of prey. It had captured a bird on the wing when it flew into a window. "It had a feather in its mouth" when it died, he tells me.

He opens another drawer. This one is full of songbirds. Purple finches, song sparrows, white-throated sparrows, and dark-eyed juncos lie belly-up in neat rows. Each bird has a white tag tied to its leg, listing the species and the place where the bird died. "These are all window kills," Klem explains.

You are probably familiar with the *thunk!* of a bird hitting a window. So is Klem. He has devoted his career to painstakingly piecing together why, when, and how birds hit windows—and how to stop them before they hit. He has studied windows at suburban homes that kill hundreds of birds in a year, and shiny glass skyscrapers that kill thousands. He has painstakingly established, step by step, why birds collide with glass and the toll that all those hits are taking on bird populations.

Professor Daniel Klem displays dozens of dead birds killed by window strikes.

Klem says it took years before conservation organizations and scientists were willing to pay attention to the dangers of windows for birds, but people are finally listening. When the international team of scientists dropped its bombshell report that the United States and Canada had lost nearly three billion birds, it pointed out that a leading cause of bird deaths in North America was collisions with windows.

A Clear Danger

THE DARK SIDE OF WINDOWS

Klem first learned about the problem of windows from a college professor. In 1974 Klem was a graduate student at Southern Illinois University at Carbondale. He was searching for a topic to research to complete his university degree. Professor William W. George shared his alarm over all the birds getting killed at windows. He suggested that window collisions might be a good subject for Klem's research.

"So the very next day, before it was light," Klem recalled, "I sat down in front of the chemistry building where [George] told me he had picked up a lot of dead birds. It was an all-glass facade [building front]. I sat there—it was January in 1974—and a mourning dove came flying through the unleafed trees and smashed right into the glass in front of me.

"I went over and picked it up and thought, *Wow, first day on the job,*" Klem said.

Next, he visited some all-glass corridors on campus. "I presumed that they were pretty hazardous," he explained. "So I looked underneath each of them. I found [bird] skeletons. I found feathers. I found imprints of birds on the glass. And I was hooked."

Klem began to gather data on window kills. He wanted to know what makes windows so dangerous and which birds were most at risk. He wrote to natural history museum directors, seeking information on window-killed birds in their collections. Klem also monitored windows at homes and university buildings around Carbondale.

But then he hit barriers of his own. At the time, many scientists believed that windows killed only a tiny number of birds. Some scientists told him that only sick and diseased birds flew into windows. Editors at top ornithological journals refused to publish Klem's research. They told him the topic was not suitable for their journals.

"I kept on scratching my head," Klem said. "I thought everything in science was suitable." He pressed on. He painstakingly pieced together that birds don't see glass. Klem learned that the head injuries they sustain when they hit glass kill some birds outright and many more later. Klem also learned that predators hunt birds that have been stunned by window collisions, adding to the death toll. As his data piled up, he became convinced that windows were a lot deadlier to birds than most people suspected. After ten years, journals finally began publishing his work. With time, other scientists started following up on his research with experiments of their own.

Klem and others have painstakingly estimated the number of birds killed at windows. Klem estimates that in the United States alone, up to a billion birds die from window collisions every year. Other studies have confirmed this number and shown that each year, another forty-two million birds die from window collisions in Canada. With findings like these, the scientific community has come around to the same dark conclusion that Klem reached: windows are major killers of birds.

 ## AN INVISIBLE KILLER

What makes birds so likely to fly into windows? Birds see the world differently than humans do. They don't realize that glass is a solid object. If a bird sees trees and sky reflected in window glass, it thinks the images in the reflection are real and flies toward them. A bird might also see the interior of a building through a window and try to take shelter inside. And then—*wham!* "Birds behave as if glass is invisible to them," Klem explained. "They're not able to tell that it's a barrier that they should avoid, and so they become victims."

Flying into a window can be as traumatic for a bird as a high-speed car crash is for a person. Some birds survive the crash, but many don't. A bird may die on impact, or it may receive life-threatening injuries. The bird may break its beak, a leg, or a wing. It may sustain a hairline fracture in its skull. The bird might look fine. It might even fly away. But it is likely to die later of its injuries.

Klem's research has shown that not all windows are equally deadly. Some windows kill no birds at all, some kill just one or two birds a year, and some kill hundreds of birds annually.

What makes a window into a bird killer? Size and location are big factors, according to Klem. Large picture windows are deadlier because of their size. So are windows in rural areas, because bird populations are most dense in these places—more birds equal more window strikes. And windows with tall vegetation in front of them kill a lot of birds, because vegetation attracts birds, drawing them close to windows.

If you like to feed birds, one of Klem's findings will be important to you. Klem has found that windows are deadlier if they have bird feeders nearby. A feeder placed 15 to 30 feet (4.6 to 9.1 m) from a window is treacherous because at this distance, a bird can fly off from the feeder and build up enough momentum to die in a window strike.

 ## BIRD SAVERS

If you think Klem's findings are bleak, also know that he has studied ways to make windows safer—and he has found lots of them. Some steps are simple. One inexpensive solution is to put decals on windows. When spaced closely together—no more than a hand's width from one another—the stickers make a window look like a solid barrier. Acopian BirdSavers are thin cords that dangle in rows outside a window, also giving birds the impression of a solid barrier. Because the cords sway pleasingly in the wind, they are also known as Zen wind curtains.

Klem has also researched how to safely feed birds, and he says the answer lies in removing bird feeders from the danger zone. If a feeder is placed more than 30 feet (9.1 m) from a window, birds are much less likely to fly into the window. Or the feeder can be placed very close to the window—less than 3 feet (0.9 m) away. At that close range, even if the bird flies into the glass, it won't have gathered enough speed to hurt itself.

Klem has even worked with glassmakers and window manufacturers to develop glass that birds can see. Bird-friendly glass is etched or marked with closely-spaced dots or lines to make it visible. The markings can also be made to be ultraviolet, making them visible to birds but not humans. But manufacturing bird-friendly glass costs time and money. Window makers will only go to the trouble of making bird-friendly products if consumers demand them.

People who feed backyard birds should hang feeders far from windows to prevent collisions.

A Clear Danger

FLYWAYS

Birds generally migrate in a north-south pattern, although birds in different places travel different routes. In the Northern Hemisphere, many birds fly north to breed in spring. In the fall, they travel south, spending the winter in warmer places. In the Southern Hemisphere (which has summer when the Northern Hemisphere has winter), birds follow an opposite pattern. In spring they fly south to breed. In fall they fly north. Some birds don't make large-scale migrations. They stay in the same area year-round but still might move with the seasons. For instance, they might spend summers on cool mountain slopes and spend winters in warmer river valleys.

When they migrate, birds usually fly along the same routes year after year. They tend to follow large landscape features such as coastlines, mountain ridges, and river valleys. The most heavily traveled routes are known as flyways. North America has four flyways: the Atlantic Flyway along the coast of the Atlantic Ocean, the Mississippi Flyway over the Mississippi River and the Midwest, the Central Flyway above the Rocky Mountains, and the Pacific Flyway along the Pacific Ocean coast. Additional flyways are found around the world.

 FLIGHT RISK

Twice a year, during spring and fall migration, Philadelphia, Pennsylvania, sits under a vast, flowing river of birds. Billions of birds move over, around, and through the city. "Philadelphia is along the Atlantic Flyway," explained Keith Russell, program manager for urban conservation for Audubon Mid-Atlantic, a regional office of the National Audubon Society. "The birds are migrating through the city in gigantic numbers." It's not the only city along a bird flyway. Major cities like Boston, Massachusetts; New York, New York; Toronto,

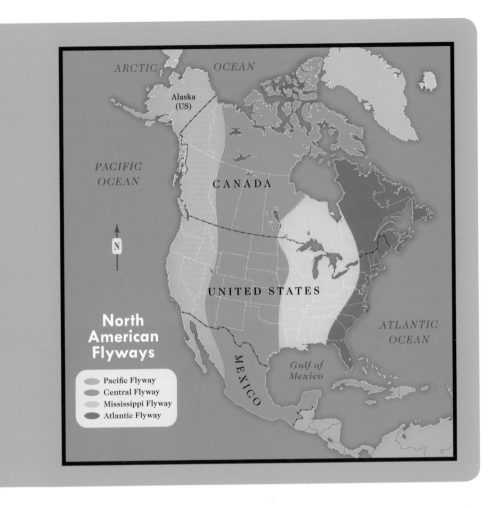

North American Flyways

Pacific Flyway
Central Flyway
Mississippi Flyway
Atlantic Flyway

Ontario; Chicago, Illinois; New Orleans, Louisiana; Los Angeles, California; and many others sit along busy flyways.

Most migratory birds travel at night. They take off just after sunset and race through the darkness. They usually fly very high in the sky, navigating by the location of stars and by energy from Earth's magnetic field (an invisible field around our planet that causes magnetic compass needles to line up north to south). When the sun rises, the birds come back to Earth and look for a stopover, a safe spot with food and shelter where they can rest during the day.

A Clear Danger

Bright city lights throw this natural cycle out of whack. Cities and suburbs are big sources of light pollution. This excessive artificial light can affect the behavior of wildlife, including birds. Most songbirds live in rural areas or forests. They are used to darkness at night. When they pass a city during migration, the bright lights attract them. They may descend into the city, searching for shelter. But the lights there can also disorient them. Some birds will endlessly circle in a beam of light, caught like a moth at a porch light. As they circle, they burn up essential energy they need to migrate. Some will drop exhausted to the ground, where they're easy prey for a predator such as a hawk or a house cat. And some will go crashing into windows.

Combine bright lights and walls of glass on towering buildings and you've got a deadly recipe for migrating birds. If lights are turned on inside a building, birds might look inside and think they have found

Tall glass buildings, such as these in New York City, can be deadly to birds.

shelter. If the building's interior is dark but the exterior is well lit, birds will see reflections in the glass and fly into windows.

"BIRDS EVERYWHERE"

To understand how lights and windows can knock out birds—and entire bird populations—with a one-two punch, consider what happened in Philadelphia on the morning of October 2, 2020. It was the peak of the fall migration. Before sunrise, Stephen Maciejewski was combing the streets of downtown Philadelphia, in search of birds killed or injured at windows. As a volunteer with Audubon Mid-Atlantic, Maciejewski was monitoring bird-window collisions at twelve downtown buildings.

That morning, pedestrians streamed past him on their way to work. Perhaps they had a cup of coffee or a smartphone in hand. Most probably did not notice the man scanning the ground for dead or dying birds. When Maciejewski found a casualty, he would kneel down, pick up the lifeless body, and slip the tiny traveler into a plastic bag, noting the time, date, and location. On a typical morning during the fall migration, he might find as many as twenty-five dead birds at the twelve buildings on his route.

But this morning was anything but typical. There were far more bodies than usual. The ground around some buildings was littered with dead and dying birds. It was a gruesome sight.

"I've never seen anything like this," he said. "There were birds everywhere."

Maciejewski had befriended several maintenance and custodial workers at his twelve buildings. They routinely helped him find dead birds on his rounds. This morning they kept bringing him bird after bird. "One guy from building maintenance dumped 75 living and dead birds in front of me," Maciejewski said.

As the sun rose and the sidewalk bustled with pedestrians,

Maciejewski hurried to collect all the victims, but the numbers kept growing. He would find fresh casualties in places he had already patrolled. There were too many birds, and he soon ran out of bags and other supplies. In desperation, he called Audubon's Keith Russell. "I immediately jumped out of bed," Russell said. "This was a big deal."

Bad weather was partly to blame. As Russell explained, "It was raining much of the night. Not only does that reduce [birds'] ability to see, but they don't want to fly in the rain." On that rainy night, clouds had formed low to the ground. This forced the birds to fly much lower than usual. The birds were racing along near the ground, partially blinded by rain, and seeking shelter. They found themselves drawn into a maze of glassy skyscrapers illuminated by bright lights. Confused by the lights and unable to see the glass, the birds began to crash into buildings and fall like rain.

That morning Maciejewski, Russell, and others collected dozens of dead black-and-white warblers, dozens of black-throated blue warblers, dozens of ovenbirds, and dozens of parula warblers. Different species of birds migrate in waves, and these were the ones that just happened to be passing through on that disastrous morning. "It takes such a toll on the populations of certain species," Russell said.

We do not know the full extent of the bird deaths that day. About four hundred birds were collected at just twelve downtown buildings, but that number was an undercount. More dead and injured birds fell onto roofs and awnings or were picked up by janitorial crews or industrial street sweepers without being counted. And some unlucky birds flew off injured, only to die somewhere else. Audubon estimates that one thousand to fifteen hundred birds died in downtown Philadelphia, with thousands more dying all over the greater metro area.

Russell said it was "sickening" to see so many dead birds. "What's even more difficult is when you see birds that are injured," he said. "They're just sitting on the ground unable to move. They're maybe

Thousands of birds crashed into buildings in Philadelphia in October 2020.

lying on their sides, or they're just barely propped up. And you say, *Wow, this could be prevented. Why does this have to happen?*"

"You just want to say, *Hey, we need to do something about this and prevent this*," Maciejewski said. "We have to bring people together to make the glass friendlier to birds. We're contributing to the extinction of American songbirds."

The tragedy on that stormy October day had a silver lining: it alerted people to the problem. Videos of the event, taken by Audubon and posted on social media, spurred people into action. The videos showed injured birds barely moving or trembling with their eyes closed. "That really is an emotionally jarring thing to see," said Russell. "They're so helpless."

Local bird organizations came together to discuss the problem. Russell penned a letter to the *Philadelphia Inquirer* newspaper, which gathered more support. The owner of the deadliest building expressed openness to finding a solution. Those conversations gave Russell hope that Philadelphia would join a movement spreading across North American cities—to make cities less dangerous for birds.

A Clear Danger

SUPERCOLLIDERS

In eastern North America, birds that are particularly vulnerable to collide with buildings include these:

- black-and-white warblers
- black-capped chickadees
- blue jays
- brown creepers
- cedar waxwings
- downy woodpeckers
- field sparrows
- gray catbirds
- house finches
- Nashville warblers
- northern cardinals
- northern flickers
- northern waterthrushes
- ovenbirds
- purple finches
- ruby-throated hummingbirds
- Swainson's thrushes
- white-breasted nuthatches
- wood thrushes
- yellow-bellied sapsuckers

What makes these birds vulnerable? Most of them are migratory, which places them at high risk for collisions as they pass through different habitats. When they pass through cities at night, for instance, they are often attracted to lit-up buildings and crash into them. The list also includes some nonmigratory birds, such as northern cardinals, blue jays, and black-capped chickadees. These birds are at higher risk because they tend to congregate at bird feeders, which are often placed at unsafe distances from windows.

LIGHTS OUT!

The first city to address the dangers of lights and glass was Toronto. The change began in 1989, when artist and gallery owner Michael Mesure heard about the problem of bird collisions in the city. Toronto sits on the northern shore of Lake Ontario. During the fall migration, the city experiences a high amount of bird traffic, as night-migrating birds pause before crossing the water on their way south. Drawn by the lights, birds drop in large numbers into the city. As a result, Toronto experiences a high number of bird-building collisions.

After learning about the problem, Mesure began patrolling the streets of Toronto's business district before daybreak, looking for dead and injured birds. Local media interviewed him, which drew more bird lovers to join him. In 1993 they formed the Fatal Light Awareness Program (FLAP) to educate people about the problem of nighttime strikes. The group soon began looking at daytime collisions as well. "The nighttime issue is bad enough," Mesure explained. "But the daytime issue statistically is far more problematic," with even more birds hitting windows. That's because most birds are more active during the day. And unlike nighttime collisions, daytime collisions don't depend on bright lights. They can happen anywhere windows are located near bird habitat. "And that's pretty much anywhere you go," said Mesure.

Mesure's art background helped him connect with the public. FLAP mailed folded paper brochures that opened into the shape of origami-style birds. Information about window collisions was printed on the birds' wings and tail. "People didn't throw [the brochures] out," Mesure said. "They kept them and showed them to their friends." In 2001 FLAP launched a "bird layout" event, an artful arrangement of all the dead birds recovered from the previous year. The event took place at the Royal Ontario Museum. "When you walk by an exhibit like that, it's just jaw-dropping," Mesure said. "People can't help but go, *How on earth have I missed this [problem]?*"

National Geographic magazine later reported on the exhibit, raising even more awareness of FLAP's efforts.

Thanks to years of FLAP's advocacy, city building owners began turning off their lights voluntarily after dark during migration season. In 2010 Toronto became the first city in the world to mandate bird-friendly features in new buildings (anything larger than a single-family house), such as markings on the outside of windows to prevent collisions. Regulations also require building owners to minimize lighting at night. Toronto has the strictest standards for bird-friendly design in the world, and FLAP is trying to get bird-friendly standards adopted into Canadian national law. "Once that happens, every building across Canada has to follow the building code," Mesure said.

The movement to make cities safer for birds is also spreading throughout the United States. In 1999 Chicago launched the

Philadelphia has joined the Lights Out program, darkening buildings at night to protect birds.

Lights Out program. From its perch at the southern end of Lake Michigan, Chicago sits along the Mississippi Flyway, another busy superhighway for birds. To protect migratory birds from collisions, almost all the major skyscrapers in Chicago turn their lights off after eleven at night during the spring and fall migrations. One follow-up study found that before the Lights Out program, a single building on the shores of Lake Michigan had killed twenty-seven thousand birds over a twenty-year period. Turning down the lights reduced bird deaths at that building by 80 percent.

Dozens of big US cities have joined the movement. In 2005 New York City, the largest city in the United States, launched its own Lights Out program. The Chrysler Building, Rockefeller Center, and other famous landmarks turn off their lights at night during peak bird migration. In 2020 the city passed the most broad-ranging bird-friendly building legislation in the United States to date. The law requires that new buildings and major renovations meet bird-friendly standards, with features such as textured glass to make windows visible to birds.

After the bird-kill event in Philadelphia, Audubon Mid-Atlantic, the city's Academy of Natural Sciences of Drexel University, the Delaware Valley Ornithological Club, and two other local Audubon chapters teamed up to tackle the problem. Together the coalition launched Philadelphia's Lights Out program in the spring of 2021. More than thirty buildings signed up to go dark during the first season. The coalition is continuing to monitor bird collisions and has invited the public to report dead or injured birds via a website (inaturalist.org). The group is also working with the Philadelphia City Council to pass bird-friendly building laws.

Keith Russell is happy to see Philadelphians taking action to make their city safer. "Birds move around," Russell said. "What happens in our cities is critical to what is happening to birds all over the continent."

HERE, KITTY, KITTY

The further disconnected we become from nature, the less we understand its complexity, its beauty, and, at times, its brutality—and we lose sight of the fact that humans are intricately part of and dependent upon the very natural systems we continue to destroy.

—American authors Peter Marra and Chris Santella

If you own a cat, it can be hard to think of your cuddly kitty as a killer. Yet all cats have deep feline instincts. They are carnivores (meat eaters) that instinctively hunt prey. When a cat sees something flutter or scurry, it can't help but stalk it. A cat will watch its prey for long periods, motionless and quiet, waiting for an opportunity to pounce. Combine deadly instincts with razor-sharp teeth and claws, and you'll quickly see why cats are extremely deadly to all kinds of birds and small animals.

KILLER INSTINCTS

Domestic cats, or house cats, first came to North America about five hundred years ago with European ships. On board, sailors kept them as pets and relied on them to kill disease-carrying rodents. Once cats reached America, their numbers soon increased. A female cat can start reproducing at around six months old. She can give birth to three litters a year, with about four to six kittens per litter. The first cats to arrive in North America quickly reproduced and spread. Cats prowl the continent.

Killing birds comes naturally to cats.

Ecologists—scientists who study relationships between living things and their environment—classify domestic cats as an invasive species, a species from elsewhere that harms native ecosystems. Cats regularly top lists of the world's most destructive invasive species. Cats are destructive partly because humans feed them. Cats generally don't have to hunt their own food, so they can survive in unusually high numbers. Being well fed in many cases, cats can become far more numerous than native predators like coyotes and mountain lions. That makes the domestic cat a kind of superpredator, an animal that is extremely deadly to wildlife.

When cats roam outdoors, lots of animals suffer, including cats. Scientific studies with kitty cams—cameras placed on free-roaming cats—have shown cats engaging in dangerous activities, such as lapping up spilled, poisonous antifreeze; dodging moving cars; and tangling with much bigger dogs. It also showed them killing plenty of birds, frogs, and field mice.

Outdoor cats take a huge toll on bird populations in rural and suburban areas. These prowling pusses eat birds that nest near the ground. They pounce on birds that land to gather nesting materials and food. They stalk young birds that are just learning to fly. Cats even prowl under windows, killing birds that lie dazed on the ground after window collisions.

In one study in Michigan, cat owners were asked to count the dead birds their cats brought home. On average, each cat killed one bird a week. Other studies have shown cats don't bring home everything they catch, so one bird a week is almost certainly an undercount.

 ## "SHOCKINGLY HIGH"

How big of a bite are cats taking out of US bird populations? Until a few years ago, no one knew. No scientific studies had been done to count the death toll on a national scale.

In 2013 scientists from two federal organizations, the Smithsonian Institution and the US Fish and Wildlife Service, joined forces to find an accurate estimate. The team sifted through earlier studies of bird kills that had focused on one city or region. They chose the most rigorous, or thorough, of those studies and multiplied the local and regional numbers to arrive at a national estimate.

The scientists reported their findings in the journal *Nature Communications*. According to their study, free-ranging cats killed 1.3 to 4 billion birds every year in the United States. The results showed that cats were slaying a sizable slice of our bird population.

Peter Marra, director of the Smithsonian Migratory Bird Center (SMBC) and an author of the report, called the death rates "shockingly high." He explained, "When we ran the model [did the math], we didn't know what to expect. We were absolutely stunned by the results." Other studies have shown that cats kill a lot of birds in Europe, Australia, and New Zealand too.

Marra's team also looked at which cats were doing the killing. They found that almost a third of the bird deaths came at the paws of our beloved pet kitties. The rest of the killings were carried out by unowned cats without homes. These homeless cats include strays that visit people's yards for an occasional dish of food and feral, or truly wild, cats with little human contact. The homeless cats were responsible for more than 70 percent of bird deaths.

The study established cats as one of the single greatest human-linked threats to birds. Like it or not, our cats are contributing to the disappearance of North American birds.

 ## CAT FIGHT!

When news outlets broke the story that cats are killing huge numbers of birds in the United States, the fur and feathers started to fly. On one side of this fight were environmentalists and bird lovers, people who believe that cats should be kept indoors to prevent them from killing birds and other wildlife. On the other side were many cat owners, people who want to give their cats a good life and believe that means plenty of outdoor time.

Reaction to the news was heated. A story in the *New York Times*, "That Cuddly Kitty Is Deadlier Than You Think," received more than a thousand online comments in the first few days. These comments pitted people urging cats be kept indoors against cat owners who wanted their precious Fluffy to roam free. Here are a few examples:

- Here in Florida we encourage our cat to roam and capture rats and mice. We encourage our cat (Rusty) to ignore cat haters who keep their cats in cages or trapped in a house with nothing but a sand box and a scratching post. I read your piece to him and he appears not the least put out and commented that cats that can't hunt are like people who prefer canned food over fresh.
- Cat owners who let their cats roam free are irresponsible and don't deserve to have cats. If you are not willing to keep your cat indoors or leash it when it goes outdoors, then you should be subject to massive and escalating fines.
- This article is ridiculous. . . . It is natural for cats to hunt. It is natural for the hawk and other raptors to hunt as well and they eat songbirds and mice too. Should we kill or make the hawks live inside? Come on.
- I'd be interested in knowing if the cat owners who are OK with their cats killing and endangering outdoor birds would stick to that survival-of-the-fittest mentality were another animal (dog, coyote, hawk etc.) to kill their cat.

Marra, who is a cat owner, teamed up with writer Chris Santella to produce *Cat Wars: The Devastating Consequences of a Cuddly Killer*. In the book, Marra and Santella call for a truce between cat lovers and bird lovers. They point out that they are all animal lovers. By tackling the problem together, we can help both cats and birds.

 WHO LET THE CATS OUT?

People on both sides of the debate agree that too many cats are homeless. How many are there? No one knows for sure. The Humane Society of the United States estimates that thirty to forty million unowned cats roam the United States, but this is just a guess. Some of these animals are feral and they shun all human contact, so they are difficult to count.

Many homeless cats are former pets that have been abandoned by people no longer able or willing to care for them. Abandoned cats often form colonies, gathering in packs, with multiple generations of related females raising their kittens together. Cat colonies form wherever food is available—either prey such as birds or food provided by people. In cities the number of cat colonies can be quite high. For example, Washington, DC, has more than three hundred colonies of unowned cats.

Cats without homes pose a great danger to birds.

How can we protect birds from the many homeless cats? Both sides agree that spaying and neutering (sterilizing) as many unowned cats as possible is a good first step to controlling the cat population. But what should we do with those cats after they have been sterilized?

Some people support trap-neuter-return (TNR) programs. In these programs, volunteers catch unowned cats and take them to veterinary clinics to be sterilized. Then the cats are returned to the outdoors. These programs have grown in popularity, with places like Houston, Texas, and San Francisco, California, embracing them. Some mainstream animal welfare organizations, including

CAT DISASTER

In August 2020, a cat crept through the ferns in the Hono o Nā Pali Natural Area Reserve on the Hawaiian island of Kaua'i. The cat prowled in a remote part of the park and in a key breeding site for endangered Hawaiian petrels. It found burrows filled with helpless petrel chicks. Their parents were far away, hunting for squid and fish with which to feed them. Left alone and unprotected, the chicks in the park were easy pickings for the free-roaming cat. Within three days, it had killed nine chicks. Scientists working in the area found the half-eaten, month-old baby birds outside of their burrows. Trail cameras revealed the culprit.

Hawaii is "the bird extinction capital of the world." according to the American Bird Conservancy. Of Hawaii's 142 endemic species— those found nowhere else in the world—95 have gone extinct.

Members of the Kaua'i Endangered Seabird Recovery Project tag a petrel, which was later killed by a cat.

Of the 47 species remaining, 33 are on the endangered species list, an official list of threatened and endangered US plants and animals. The precarious status of Hawaii's birds is due, at least in part, to cats.

Birds arrived in Hawaii around ten million years ago, when the island chain was still forming from erupting volcanoes. Those early birds flew to the remote Pacific islands or were blown there by storms. The islands offered them diverse habitats—from mountain peaks to beaches—and a wide variety of foods, including insects, fruits, and seeds. The birds multiplied and evolved into diverse species.

Polynesian settlers, who came to Hawaii by canoe around 500 CE, brought the first non-native animals to the islands. These animals included rats, which preyed on the native birds. Cats first came to the islands with European explorers in 1778. They multiplied and quickly became the island chain's most widespread predator.

In modern times, Hawaii's birds are threatened by habitat loss, light pollution, and diseases. But the two million cats that prowl the eight Hawaiian Islands pose the worst danger to Hawaii's bird populations. State officials say the island is facing a full-blown "cat crisis."

Conservationists haven't given up on saving Hawaii's remaining birds. Wildlife officials are trapping homeless cats, sterilizing them, and encouraging adoption.

Conservation departments are also building predator-proof fences around key bird habitat areas. The tall steel fences keep out animals as small as mice. The fences have rolled tops that prevent cats and other predators from climbing over. People can access the fenced-in areas through locked gates.

the Humane Society and the American Society for the Prevention of Cruelty to Animals, also support TNR.

Supporters of TNR believe it is a humane solution to cat overpopulation. They say it improves the lives of homeless cats, because veterinary staff also vaccinate cats against diseases when they are brought in to be sterilized. Supporters point out that because the sterilized cats can no longer reproduce, they no longer contribute to the problem of cat overpopulation. With enough sterilized cats, the homeless cat population will gradually decrease, supporters say. But scientific studies put this claim in doubt. The problem, studies show, is that you've got to catch and sterilize nearly all the stray cats in a colony before you make a dent in the rate of reproduction. And cat colonies are constantly receiving an influx of newly abandoned, unsterilized cats, so it's nearly impossible for TNR teams to keep up.

All this is bad news for birds. Between the millions of beloved and cared-for cats allowed to prowl outdoors and the millions of wild, homeless cats roaming streets, woods, and meadows, birds remain in the crosshairs.

HAPPY CATS, HEALTHY BIRDS

How can we solve cat overpopulation? How can we protect both cats and birds? Nearly everyone agrees on a few things: We should spay and neuter as many cats as we can. We should get as many cats adopted and into safe homes as possible. And pet cats should always be kept indoors. This last step involves educating cat owners, and this work has already begun. The American Bird Conservancy runs a Cats Indoors public awareness campaign. This program communicates to cat owners the perils of free-roaming cats.

Another innovative solution is fencing off cat colonies. In central California, an enclosed 12-acre (4.9 ha) sanctuary provides a home to more than seven hundred cats. When cats are admitted to the sanctuary, they are sterilized and vaccinated. Outbuildings on the property give cats places to take shelter. All the cats are available for adoption, and nearly five hundred are placed with human families each year.

Marra and Santella call for strict no-outdoor-cat policies in places that are critical to bird conservation. These places include Cape May, New Jersey, and Galveston, Texas, both stopovers for migrating birds. Another site needing protection is Hawaii. The Hawaiian Islands are home to thirty-three endangered bird species that are found nowhere else in the world. In such important bird spaces, Marra and Santella argue for zero tolerance: no free-ranging cats. "If the animals are trapped, they must be removed from the area and not returned," they write.

The cat problem is a human problem—one we have made and one we can solve. If we work together, we can solve the problem quickly and help birds bounce back. As Marra and Santella remind us, "Nature is resilient once given a chance."

HAWK WATCHING

It was a spring without voices. On the mornings that had once throbbed with the dawn chorus of robins, catbirds, doves, jays, wrens, and scores of other bird voices there was now no sound; only silence lay over the fields and woods and marsh.

—American biologist Rachel Carson

The blue autumn sky arched high over a rocky overlook in the Appalachian Mountains of eastern Pennsylvania. The birds just kept coming: ospreys flying on M-shaped wings, bald eagles, red-tailed hawks, sharp-shinned hawks, broad-winged hawks, and kestrels.

I encountered the birds on Hawk Mountain, home to the longest-running hawk watch in North America. Along with David Barber, a biologist at Hawk Mountain Sanctuary, and a few dozen other bird-watchers, I was seated on a big pile of rocks high on the mountain. This rocky overlook offers a vantage point across the

A red-tailed hawk in flight

spine of the mountain ridge and the valley below, where highways run past houses, barns, and farm fields. In a fenced-off area, counter Greg George of Hawk Mountain Sanctuary and volunteer Breanna Bennett wore matching khaki-colored vests and waited with an iPad, binoculars, and a small, high-powered telescope. Throughout the fall migration, paid counters and volunteers staff the overlook every day from sunrise to sundown.

"Hawk watching" really means "raptor watching." Raptors—birds of prey—are a large and diverse group. They include hawks, eagles, falcons, kites, vultures, ospreys, and owls.

RIVER OF RAPTORS

Every fall, up to ten million raptors migrate between their breeding grounds in the United States and Canada and their wintering grounds in Mexico, Central America, and South America. Flying from Canada to South America is an exhausting undertaking. So raptors do whatever they can to save energy. One trick is to fly along mountain chains.

Hawk Mountain sits along the Kittatinny Ridge, which stretches for 300 miles (483 km) and is the easternmost ridge of the Appalachian Mountains. When brisk autumn winds strike the ridge, they blow upward along the side of the mountain. These rising winds, or updrafts, give a helpful lift to passing raptors. Instead of flapping, the birds can spread their wings and

Bird-watchers at Hawk Mountain scan the skies for raptors.

soar. On days with favorable flying conditions, raptors from all over the northeastern United States and Canada pour southward along the stony spine of the Kittatinny. Broad-winged hawks, red-tailed hawks, sharp-shinned hawks, bald eagles, golden eagles, ospreys, falcons, merlins, kites, and kestrels all funnel past the rocky overlook. From September through November, an average of eighteen thousand birds soar past this spot.

On my day at Hawk Mountain, the show began with a hawk. Bennett spotted it first, calling that it was coming right toward us. Binoculars lifted in unison. Barber identified it as a broad-winged hawk, a crow-sized bird with a black-and-white tail. The hawk circled, riding a column of warm, rising air, going higher and higher. As the hawk approached the lookout, it banked steeply and soared over the valley.

Next came an osprey. Then a sharp-shinned hawk. Then two bald eagles, young ones with chocolate-brown bodies and brown-and-white wings.

Bennett called out each bird, and George recorded it on the iPad. As the counts at Hawk Mountain show, at a time when many North American birds are in decline, raptors are a bright exception. Most are thriving. But that wasn't always true.

SHOOTING GALLERY

In the early 1900s, on the rocky overlook at Hawk Mountain, men with shotguns and rifles stood around on many autumn days, cracking jokes and commenting on the weather. When raptors came soaring past, the men raised their guns and took aim. Then came the crack of gunfire, the smell of smoke, a few seconds of silence, and the crackling of birds crashing on dried leaves. Some of the birds were already dead. Others would die slowly of their injuries. All were left to rot on the forest floor.

Americans of earlier eras despised birds of prey. Farmers hated them for killing chickens. Bird lovers hated them for killing songbirds. Hunters didn't like birds of prey killing grouse and pheasants, since the hunters wanted to shoot these birds for sport.

By shooting birds of prey, people believed they were giving other living things a helping hand.

Shooting raptors became a major activity after the Civil War (1861–1865). Shooting was heaviest along migration flyways, at places like New Jersey's Cape May and Pennsylvania's Kittatinny Ridge. Some states encouraged the shooting. Alabama and Virginia launched anti-hawk campaigns, enlisting hunters and conservation clubs to help wipe out certain species. Pennsylvania enacted a bounty, paying hunters who shot certain birds of prey. By the 1920s, raptor populations in the eastern United States were in free fall.

Even prominent scientists and conservationists supported the killing. In 1931 William T. Hornaday, a pioneer in the wildlife conservation movement, said about the peregrine falcon, "Each bird of this species deserves treatment with a choke-bore gun. First shoot the male and female, then collect the nest, the young or the eggs, whichever may be present. They all look best in [museum] collections."

Not everyone agreed though. As raptor numbers fell lower and lower, public sentiment made a slow U-turn.

 ## A SANCTUARY FOR HAWKS

In 1932 Richard Pough, a recent college graduate and a bird lover, set out to see the Hawk Mountain shooting ground. He arrived on the mountain on a day crowded with hunters and witnessed the shooting for himself. On another visit, when no hunters were present, he found the ground littered with dead birds. He lined up hundreds of carcasses and took pictures of them.

In 1933 Pough spoke out against hawk hunting at a New York City meeting of conservationists. He showed his pictures. A bird-loving woman in the audience was moved by the photos. Rosalie Edge had been involved in the movement to win voting rights for US women. She was used to fighting for causes she believed in. Determined to end the slaughter, she raised funds to buy nearly 1,400 acres (567 ha) of land on the Kittatinny Ridge. In 1934 she established Hawk Mountain Sanctuary as the world's first sanctuary for birds of prey.

Edge hired ornithologist Maurice Broun to count all the raptors migrating past the lookout, thus launching North America's first official hawk watch. Broun's wife, Irma, acted as gatekeeper, turning back hunters so that her husband could count the birds gliding by.

Hawk Mountain Sanctuary helped swing public opinion. Bird lovers flocked to the mountain to witness the migration. By the late 1940s, the mountain was hosting tens of thousands of visitors every year. As support for raptors grew, widespread shooting faded, and many states passed laws protecting birds of prey.

One early visitor to Hawk Mountain was US marine biologist Rachel Carson. She arrived at the rocky overlook on a cold, windy, gray day in October 1945. In her field notes, she mused about being at the bottom of "an ocean of air on which the hawks are sailing."

 ## TOO GOOD TO BE TRUE?

By chance, Carson's visit to Hawk Mountain took place the same year that a new "wonder chemical" was introduced in the United States. Carson would later use official bird counts from Hawk Mountain to make the case that this new chemical was bad for birds, especially birds of prey. The wonder chemical was dichloro-diphenyl-trichloroethane. The public came to know it by its easy-to-pronounce abbreviation DDT.

DDT was a new kind of insecticide—a chemical that kills insects.

After its introduction in 1945, DDT was widely used. Farmers sprayed it on their fields to kill insects that damaged crops. Public health agencies sprayed it in marshes to kill mosquitoes that carried disease. Rangers sprayed it in forests to kill insects that hurt trees. Manufacturers added it to plastic kitchen shelf linings to keep bugs away from food and embedded it in carpets to kill fleas that could harm pets.

DDT seemed almost too good to be true. It was long-lasting, inexpensive, and killed a broad range of insects. What's more, it seemed less toxic to people and wildlife than previous pesticides. So revered was DDT that its inventor and proponent, Swiss chemist Paul Müller, won the 1948 Nobel Prize in Physiology or Medicine. The prize goes to those whose work benefits humankind, and DDT was then seen as a great benefit to people.

But the widespread use of DDT soon raised alarms. Studies showed that it causes cancer in people. It is toxic to a wide range of water-dwelling animals, such as crayfish and many species of fish. It also harms birds, especially birds of prey. DDT interferes with a bird's ability to process the mineral calcium, the building material of eggshells. Birds exposed to DDT lay eggs with thin, deformed shells. When the mother or father bird incubates, or sits on, an egg in the nest, the eggshell breaks, killing the developing bird inside.

Although DDT was harmful to many living things, raptors were at high risk because DDT moved up the food chain. For instance, small fish and tiny animals called zooplankton absorbed DDT from the water. When larger fish ate smaller fish or zooplankton, more DDT built up in their bodies. Peregrine falcons tended to have high levels because they hunted in DDT-soaked farm fields, preying on the small birds that had fed on DDT-contaminated insects. Eagles and ospreys had high levels because they fished in waterways contaminated with DDT from farms and forests.

Very soon after DDT's introduction, raptor populations began to plummet. By the 1950s, bald eagles, ospreys, and peregrine falcons were in a deep dive, along with Cooper's hawks, sharp-shinned hawks, merlins, black vultures, and turkey vultures. By 1970 only a few hundred bald eagle pairs (male and female mates) remained in the United States. The bald eagle, the peregrine falcon, and many other raptors landed on the endangered species list.

"A SPRING WITHOUT VOICES"

The public first became aware of the DDT problem with Rachel Carson's 1962 book, *Silent Spring*. She laid out the perils of DDT and similar pesticides and warned of "a spring without voices," one in which no birds sang. She cited as evidence the severe drop in raptors observed in the annual counts at Hawk Mountain.

Carson's book shook the American public and, along with the growing environmental movement, helped pave the way for the 1970 creation of the US Environmental Protection Agency (EPA). This federal agency is tasked with protecting human health and the environment. In 1972 both the United States and Canada banned most uses of DDT. European countries also banned DDT, starting with Hungary in 1968. In 2001, 170 countries signed a treaty restricting the use of DDT to emergency insect control—for example, to kill mosquitoes that spread deadly malaria.

The ban on DDT worked. By the mid-1980s, many raptors were rebounding from their lows in the DDT era. And in the 2020s, most North American raptors are thriving. More than thirty thousand eagle pairs are living in the wild.

Raptors have become so common that some people again consider them a problem. In 1999 the Pennsylvania Game Commission held hearings on a proposal to kill red-tailed hawks and great horned owls on state-owned lands. Why? To help populations of

Biologist Rachel Carson works at her home laboratory in Maine, 1962.

game birds such as ring-necked pheasants, because raptors were hunting the pheasants before human sport hunters could kill them. Although some people spoke up in opposition and the proposal was withdrawn, many others supported the idea. In 2002 the US Fish and Wildlife Service allowed the killing of four hundred vultures in the US Southeast after complaints of vultures preying on newborn calves and lambs. And Hawk Mountain Sanctuary regularly fields phone calls from bird lovers outraged over hawks preying on songbirds at their feeders.

What is the solution to keeping birds of prey (and all birds) abundant and thriving? As raptors have revealed, removing the threats they face is key. So is habitat conservation and long-term monitoring, helping us understand population trends and head off any new and emerging threats. But education is also enormously important. When people can see raptors up close at places like Hawk Mountain Sanctuary, their experiences build support for conserving birds.

6

FOOD CHAINS AND BIRD BRAINS

The Bobolink is gone—
The Rowdy of the Meadow—
And no one swaggers now but me

—American poet Emily Dickinson

On a spring morning, Christy Morrissey drives her pickup truck through rolling farmland in Saskatchewan, Canada. Fields of canola plants stretch as far as the eye can see. Shallow ponds called potholes pockmark the land. Each pond is fringed with cattails and provides a home for ducks. Tree swallows swoop and dive over the ponds, feeding on flying insects.

Morrissey's truck crunches to a stop along a gravel road. She grabs a cotton bag and creeps toward a wooden birdhouse nailed to a post. Working quickly, she claps the bag over the round birdhouse entrance hole and carefully opens a door on the side. She reaches her hand inside, gropes for a moment, and says, "Got her!" She tips a glossy, greenish-blue tree swallow into the bag.

The swallow is a mother bird with a nest full of eggs. Morrissey pulls the bird from the bag, reads a metal band on her tiny leg, and carefully weighs and measures the bird before releasing her. Morrissey will return in a few weeks to collect blood and feather samples from the mother's chicks. That data, together with data on nearly five hundred other tree swallows, helps her assess the birds' stress levels, what they're eating, and their overall health.

Morrissey is a wildlife ecotoxicologist, a scientist who studies the effects of toxic substances on wildlife and the environment. She is comparing the health of birds that live in native grasslands with those that live near big, industrial farms. Her work probes a question that worries ornithologists: Are birds being harmed by the way we grow food?

Scientists Anson Main and Christy Morrissey test pond water for pesticides in Saskatchewan, Canada.

Food Chains and Bird Brains

A NEW PESTICIDE PROBLEM

Before the twentieth century, a sea of grass covered the middle of
North America. Prairie grasses waved, green and gold, from Canada
to Mexico. Those grasslands were alive with the songs of birds. If
you walked through a western prairie in spring, you might hear
the flutelike notes of a western meadowlark. Farther east, you'd
hear the clear whistling song of its cousin, the eastern meadowlark.
In either place, you might see the fluttering flight and hear the
spirited babble of a bobolink—poet Emily Dickinson's "Rowdy of
the Meadow."

But our grasslands are in trouble, as are the birds that depend
on them. Most of their grassy habitat has been plowed or paved,
converted to farmland or sprawling suburbs. Less than 40 percent
of our native grasslands remain. Much of what is left has been
carved up into tiny patches, where birds struggle to find enough
food, to find mates, and to find places to raise their young. As
a result, grassland bird populations are dropping like a stone.
Since 1970 western meadowlarks are down 48 percent, eastern
meadowlarks are down 73 percent, and bobolinks are down
71 percent. Overall, nearly three-quarters of all grassland bird
species are in decline.

With much of their habitat gone, grassland birds have had to
make do. Many nest on farms—in hay fields, grain fields, pastures,
idled cropland, and the unplowed strips of land that line farm fields.
But farms have changed, in ways that threaten the birds that live
there. Big corporations have bought up many small family farms
and merged small fields into enormous ones. Places where birds
previously nested, such as hedgerows between fields, have been
cut down. Big corporate farms often use monocropping, or planting
just one crop over a large area, which depletes nutrients in the
soil. Corporate farmers are also using more chemicals: chemical
fertilizers to help plants grow, chemical herbicides to kill weeds,

chemical insecticides to kill insects, and chemical fungicides to prevent crop diseases.

This suite of changes is known as agricultural intensification, and studies in Europe and North America show that birds are disappearing faster in areas where agriculture has intensified. The prime culprit? Pesticides. In Europe and North America, studies have linked large-scale pesticide use to plummeting bird populations.

In 1962, when Rachel Carson published *Silent Spring*, the most widely used pesticides in the world were organochlorines, a group that includes DDT. After bans on DDT and other organochlorines in the 1970s, new pesticides came into use. But many of the newer chemicals also proved dangerous to people and animals. Some were later banned or restricted by the EPA, as well as by government agencies in Canada and Europe.

On large industrial farms, workers spray chemical pesticides and herbicides to kill insects and weeds.

Food Chains and Bird Brains

Chemical companies and the EPA searched for new insecticides that would be safer for people and wildlife. In the 1990s, scientists invented a new class of pesticides based on nicotine, a natural insecticide made by tobacco plants. The new pesticides are neonicotinoids, or neonics for short.

Like many insecticides, neonics are neurotoxins. They work by damaging an insect's central nervous system. Plants treated with neonics absorb the chemicals into their roots, stems, and leaves. If an insect munches any part of a neonic-treated plant, it gets a dose of pesticide and drops dead.

Neonics have become the most popular pesticides in North America and the world. They are most commonly applied not to growing plants but on seeds before planting. In the United States, almost all corn seeds are coated with neonics. So are about half of all soybeans and nearly all cotton seeds. In just these three crops alone, US farmers plant neonic-treated seeds on at least 150 million acres (61 million ha)—an area roughly the size of Texas. In Canada, neonic seeds are planted on 44 percent of farmland, including 21 million acres (8.5 million ha) of canola seeds.

The colorful coating on these seeds contains a chemical pesticide.

Neonics are designed to kill insects that feast on crops, but they end up spreading into the surrounding environment. Growing plants absorb only a tiny fraction of the neonics from their coated seeds. The rest of the chemicals are washed away with melting snow and falling rain. They enter rivers, lakes, streams, and ponds. There, they poison insects that hatch in the water and the birds that eat those insects.

PRAIRIE FOOD CHAINS

In the pond-studded canola fields where tree swallows swoop, Christy Morrissey investigated whether neonics were affecting songbirds through their food chain. She and her team worked in painstaking fashion, focusing on one part of the food chain at a time.

They began by testing whether neonics were washing into ponds in farm fields. And, indeed, the chemicals were there. "Most of those wetlands are contaminated with neonics," Morrissey said. Her team detected neonics in up to 91 percent of the ponds they studied.

Next, they investigated whether the pesticides could be harming midges. These tiny flies make up 70 to 80 percent of insect life in the ponds. Adult midges lay their eggs in the water, their larvae (immature offspring) develop underwater, and the flying adults emerge into the air, where they are gobbled up by tree swallows. Morrissey's team discovered that midges are highly sensitive to neonics in the water. The chemicals can kill the larvae outright or can have subtler effects, like causing midges to be smaller and to emerge earlier from the ponds. This early emergence of midges can be bad for migratory birds, whose spring arrival at their breeding grounds is timed to coincide with peak insect abundance. "Birds are highly reliant on those insects," Morrissey explained.

Food Chains and Bird Brains

Because of that reliance, the team next looked at how changes to midges could be affecting tree swallows. They compared the health of swallows that live near neonic-treated fields with those that live in native grasslands. They found that swallows near farms have smaller chicks and spend more time away from the nest. Both are ominous signs. They suggest that the birds are struggling to find enough insects to eat.

BYE-BYE, BUGS

Insects dominate life on Earth. Around 80 percent of all animals on the planet are insects. On any given day, ten quintillion insects (ten followed by thirty zeros) fly, swim, or crawl around our planet. They play an outsize role in maintaining the world's ecosystems. For example, when bees fly from flower to flower, they pollinate most of the world's flowering plants—including those we rely on for food. Insects also serve as food for a huge number of other animals, including birds. US biologist E. O. Wilson observed that "if we were to wipe out insects alone, just that group alone, on this planet the rest of life and humanity with it would mostly disappear from the land. And within a few months."

In 2017 scientists reported in the online journal PLOS One that the mass of insects at a nature preserve in Germany had declined 76 percent from 1989 to 2016. That finding made headlines around the world. The *New York Times* warned of an "insect Armageddon [massive destruction]."

The finding of massive insect losses in a German nature preserve prompted entomologists, or insect scientists, to take a close look at insect populations in other locations. Insect declines appear to be widespread, including in North America. In the White Mountains of New Hampshire, the number of forest beetles has fallen 80 percent since the 1970s. In the upper midwestern United States, mayfly numbers have dropped by more than half since 2012.

Why are insects disappearing? Entomologists are scrambling to find answers. Loss of habitat could be playing a role. Without their natural homes, some insects cannot find enough food, water, and shelter. Due to increased carbon dioxide in the air, Earth's temperature has risen by more than 1°F (0.6°C) since the early twentieth century. Higher temperatures have increased the likelihood and intensity of extreme weather, and many insect species, particularly those that live near the equator, are not adapted to wild swings in weather. So climate change could be a culprit in insect deaths. Another likely culprit is pesticides, such as the neonics showing up in ponds in Saskatchewan's canola fields.

Whatever the cause, insect declines can hit birds hard, especially birds with insect-heavy diets such as swallows, swifts, warblers, and woodpeckers. Lack of food can reduce chick survival and make reproduction more difficult. Birds won't lay eggs if they don't have enough food. If they do manage to reproduce successfully, they have to work harder to find food for their nestlings. Morrissey is seeing tree swallows with smaller chicks and noticing parents spending more time away from the nest looking for food, leaving them less time and energy to devote to guarding their chicks from predators.

But the food chain is only one way that pesticides could be affecting birds. Morrissey's team has also gathered evidence that pesticides could be harming birds directly—by affecting their brains and messing up their ability to migrate.

 ## THE LATE BIRD

In spring 2017, Morrisey's colleague Margaret Eng set up a mobile laboratory at Long Point Bird Observatory in Southern Ontario. The observatory sits on a narrow peninsula jutting out into Lake Erie and is home to a long-running bird-tracking program. The

Researchers study white-crowned sparrows to see how neonicotinoids affect their health and behavior.

program traps migrating birds by snaring them in mist nets—fine-meshed nets that the birds can't see. Ornithologists and volunteers carefully untangle each trapped bird and wrap a lightweight metal band around its leg before releasing it to finish its migration. The band is stamped with a number and gives details on the species as well as the date and place of the bird's capture. Bands allow observers to see which birds have been captured before and can help researchers keep track of birds as they migrate.

To help Eng with her project, bird banders at Long Point didn't release any white-crowned sparrows they caught in their mist nets. Instead, they handed the sparrows off to Eng. White-crowned sparrows have pale gray bodies and bold black-and-white stripes on their heads. They breed in brushy habitat in Canadian forests and stop in farmland during their spring migration, hopping on the ground in search of seeds. Eng wanted to learn what happens when birds such as these ingest small amounts of neonics.

Inside her mobile laboratory, Eng weighed each bird and then slipped it into a clear plastic tube. She inserted the tube into a dishwasher-sized quantitative magnetic resonance machine. The device scanned the sparrow's body and measured the amount of water, lean mass (bone and muscle), and fat it contained. She then separated the sparrows into two groups. With one group, she used a syringe to feed each bird a tiny dose of neonics. A second group, the control group, received no pesticide.

Eng put the birds in cages. She gave them food and water, and they could eat and drink as much as they wanted. Six hours later, she again weighed each bird and scanned it in the quantitative magnetic resonance machine.

In that short span of time, she found that the pesticide-dosed birds lost on average of 6.5 percent of their body mass. That's like a 130-pound (59 kg) teenager losing 8.5 pounds (3.9 kg) between lunch and dinner. Eng discovered that almost all the lost weight was from body fat, a bird's main fuel for migrating. But the control birds didn't lose body mass or body fat.

Why had the birds lost weight after eating a bit of pesticide? By weighing the amount of food in the birds' cages, Eng discovered that the dosed birds had stopped eating. In people, nicotine suppresses appetite. Neonics were doing something similar to birds.

Eng wondered how this exposure would affect birds in the real world. After ensuring that each bird was "still hopping around and happy," she glued a tiny tracking device onto each bird's back. The devices emitted radio pulses that could be picked up by a network of receiving stations.

Then Eng opened the cage doors and let the birds fly out. As she expected, the control birds flew north, continuing their spring migration at once. But the dosed birds were not in a hurry. They lingered on the Long Point Peninsula for three and half days before continuing their journey. Why? Eng explains that the dosed birds were instinctively waiting until the intoxicating effects of the pesticide had worn off and until they had eaten enough to recover the fat they use for fuel. She says that birds lose the urge to migrate if they don't have enough body fat. "How much fuel birds have is a really important part of their migration," she explains.

Food Chains and Bird Brains

Morrissey agrees. "That fat reserve that they are losing is everything they need to fly north," she says. "It's like we just punctured the gas tank."

Eng's study, published in *Science* in 2019, showed that low doses of neonics had the potential to disrupt bird migration. And even a small delay can be a problem, because that could cause birds to arrive late to their breeding grounds and interfere with their ability to reproduce—because the early birds don't just get the worms. They get the healthiest mates and the first choice of nesting sites. The late bird loses out and may even skip breeding altogether.

"PLAYING WHACK-A-MOLE"

As evidence grows that neonics could be harming birds and other wildlife, some US states have restricted them, and the EPA is reviewing their use. But Morrissey doesn't believe a ban on neonics is the answer to protecting wildlife. "The problem is so much bigger than neonicotinoids," she says. "There are thousands of chemicals that are being used." This long list of chemicals used on crops includes many types of insecticides, herbicides, and fungicides. And Morrissey points out that chemical companies are already producing new insecticides to replace neonics in the event of a government ban. "It's like we're playing whack-a-mole," she says. "Knock one [pesticide] out and a whole bunch more pop up."

She believes the solution lies in a bigger change, one that recognizes that farms are deeply connected to the natural world. "We have to change the whole agricultural system and how we grow food in order to release our reliance on chemicals so we aren't poisoning the landscape." One step involves education. Morrissey reports that many farmers aren't even aware that they are using neonics. They simply buy the seeds available from seed companies and plant them in the ground. Morrissey is talking with farmers about how they can balance the demands of growing food while protecting birds and other wildlife. She finds that many farmers are open to solutions. "I think there are many concerned farmers out there," she said. "They also notice that there are fewer birds and so they're asking questions."

Sustainable farming practices can be the solution. One sustainable approach is organic farming. Organic farmers protect soil and water to protect the environment and people. They follow strict rules, including using no synthetic (human-made) pesticides, herbicides, and fertilizers. Rather than monocropping, they rotate many crops through their fields to replenish the soil. To control insect pests, they welcome natural predators, such as birds, bats, and predatory insects. Instead of using chemical fertilizers, they enrich the soil with compost (decaying plant matter) and animal manures. They grow crops such as clover and grasses and then plow them back into the soil, which also enriches it.

Another sustainable practice is integrated pest management. Farmers who practice this follow a series of steps to deal with insect

A farmer inspects spinach at an organic farm.

Food Chains and Bird Brains

MADE IN THE SHADE

Many of the colorful birds that sing to Americans and Canadians in the spring spend their winters in Mexico, Central America, and South America. These places are also home to many coffee growers. Coffee plantations can provide good or bad habitat for birds. It all depends on how the coffee is grown.

Coffee originated in Africa. Dutch settlers brought the plant to the Western Hemisphere in the 1700s. Coffee production began on the island of Martinique and spread throughout the Caribbean Sea, South America, and Central America. Farmers grew coffee plants on the forest floor, under a dense canopy of trees. This traditional growing method has plenty of advantages. The trees and plants provide good habitat for birds of the forest. Chemical fertilizers aren't needed because fallen tree leaves rotting on the ground are full of nutrients to feed the coffee bushes. Pesticides aren't needed because there are plenty of birds to eat insect pests. And coffee grown this way produces a high-quality, rich-tasting brew.

Almost all coffee was grown in the shade until 1972, when scientists developed a coffee plant that could grow in full sun. The main advantage of sun-grown coffee is that it produces higher yields—more coffee per acre. But this method has many disadvantages, including being bad for birds. Forests must be

pests, trying the least toxic methods first before resorting to more toxic ones. One safe method is to lure the insects into traps using harmless chemicals and then physically removing the traps from the farm. Synthetic pesticides are used only as a last resort, when safer methods have failed.

Sustainable farming also includes restoring lost habitat to farms. Unplowed fields, patches of grassland, and strips of sunflowers planted among crops all give grassland birds places to nest and cut down, and chemical fertilizers and pesticides must be used to make up for the rotting leaves, insect-eating birds, and other natural advantages of shade-grown coffee. In some places, sun-grown coffee fields dominate the land—leaving no habitat for birds. At least forty-two species of North American songbirds spend winters in coffee farms, and twenty-two of those species are in decline.

Many environmentally conscious coffee drinkers choose brews that are certified as shade-grown. The Bird Friendly program, administered by the Smithsonian Migratory Bird Center, is considered the gold standard for shade-grown coffee. The center has done much of the research connecting birds and coffee plantations. To qualify for its Bird Friendly seal, coffee must be grown organically and meet strict shade-cover requirements. Another group that certifies shade-grown coffee is the Rainforest Alliance, an international group working to protect forests and fight climate change.

Shade-grown coffee has become easier to find at coffee shops and on store shelves, including Starbucks and Whole Foods. Shade-grown coffee costs more that sun-grown coffee, but it supports both birds and the farmers who grow it.

forage. And these restored habitat areas can have big benefits for farms. Habitat strips and patches increase the diversity and abundance of bird species on farms. And farms with more birds tend to have fewer insect pests, because birds eat the pests.

Morrissey believes that moving toward more sustainable ways of growing food is not only necessary, it is completely achievable. "We can manipulate agriculture in any way we want," she said. "This is something we can fix, particularly if there's enough outcry."

SEA CHANGE

How inappropriate to call this planet Earth when it is clearly Ocean.

—British author Arthur C. Clarke

On New Year's Day in 2016, David Irons was driving past a beach in Whittier, Alaska. He spotted a long line of white lumps near the water's edge. Irons, a recently retired seabird biologist with the US Fish and Wildlife Service, thought the lumps might be melting snow. But when he stopped for a closer look, he discovered they were the white-and-gray bodies of common murres. The dead seabirds had washed ashore. "It was pretty horrifying," Irons said.

Aided by his wife, son, and a friend, he counted the dead birds on the beach. The body count climbed quickly. Along a beach about a mile (1.6 km) long, they found eight thousand dead common murres.

When dead common murres washed up on Pacific beaches, scientists searched for the cause.

Irons wasn't the only one to find dead birds. Since 2015 people had been reporting dead murres on Pacific beaches from California north to Alaska.

Soon researchers got involved. First, they tallied all the reports of dead birds from the public. Then they contacted wildlife rehabilitation centers and learned about more dead birds. Finally, professional biologists and trained citizen scientists fanned out to remote Alaskan beaches to find more bodies. Researchers estimate that between the summer of 2015 and the spring of 2016, sixty-two thousand dead murres had washed ashore from California to the Bering Sea. Most had been found in the Gulf of Alaska.

What had happened to the murres? Had they been devastated by a disease? Had they been poisoned? What had killed so many birds?

 ## A TRAIL OF DEAD BIRDS

Common murres look like skinny penguins. They live most of their lives over coastal waters. They soar above the waves and dive deep for fish, paddling with their long, flipperlike wings.

Sea Change

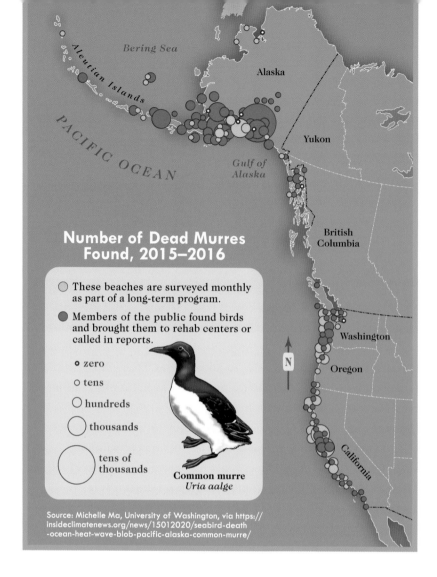

Number of Dead Murres Found, 2015–2016

○ These beaches are surveyed monthly as part of a long-term program.

● Members of the public found birds and brought them to rehab centers or called in reports.

∘ zero

○ tens

○ hundreds

○ thousands

○ tens of thousands

Common murre
Uria aalge

Source: Michelle Ma, University of Washington, via https://insideclimatenews.org/news/15012020/seabird-death-ocean-heat-wave-blob-pacific-alaska-common-murre/

Murres come to land in spring to breed. In western North America, they nest on cliffs along the Pacific coast from California to Alaska. Common murres also nest along the Atlantic coast from Maine to Canada and on coasts across northern Europe and Russia. Their colonies are crowded, noisy places, with birds sometimes nesting shoulder to shoulder.

Common murres are one of North America's most abundant seabirds. About 6 million breed in western North America, part of a worldwide population of 13 to 20.7 million birds. They are a hardy species, known for adaptability and resilience. Why then had so many of them died in 2015 and 2016?

As reports of dead murres rolled in, John Piatt, a marine ecologist with the US Geological Survey, grew perplexed. "Murres are the ultimate predator—they're extremely well adapted. They can dive to 200 meters [656 feet]. . . . And they're the fastest flying seabird." Yet dead murres were washing ashore up and down the western coast.

To find out why, Piatt teamed up with more than twenty researchers from US government agencies, universities, and conservation groups. They published their findings in a 2020 report in the journal PLOS One.

They reported that the true death toll was much higher than the number of birds that had washed in with the tide. When birds die at sea, only a fraction of the bodies reach land. The team estimated that as many as 1.2 million common murres had died in 2015 and 2016, up to 20 percent of the total population from California to Alaska.

Researchers shipped carcasses to the Alaska Science Center in Anchorage and the National Wildlife Health Center in Madison, Wisconsin. "They did all sorts of analyses for viral and bacterial diseases, toxins in the tissues," said Julia Parrish, a University of Washington biologist who was involved in the study. The autopsies turned up no toxins, no diseases, and no parasites. "Nothing we can hang our hat on," she said.

But one finding was consistent. "There was lots of emaciation," Parrish said. Many of the birds had died of starvation. Others, weakened from hunger, had probably perished in storms.

Common murres catch finger-length fish to feed themselves and their young. They can also feed on shrimplike krill. They need to eat 60 to 120 small, fatty fish every day. If they can't find enough food, they will burn through the fat they have stored in their bodies. Once that fat is gone, their bodies will begin using their muscles as fuel.

It doesn't take long for starvation to set in. As Piatt's team wrote in the study, "If murres can't fully meet [their] food demand every day, they lose body condition [health] quickly and jeopardize survival. If they can't find any food for 3–5 days, they will die of starvation."

But murres can fly up to 60 miles (97 km) per hour and travel miles to find food. They won't starve unless fish are scarce over a huge area of ocean. That made Piatt wonder, "What could account for a decline in the food supply from California to the Bering Sea [west of Alaska] all at the same time?" Piatt's team soon turned their attention to a prime suspect: "the Blob."

 ## THE BLOB THAT COOKED THE PACIFIC

It sounds like something out of a horror movie—an alien life-form that digests everything in its path. But the Blob was not an alien. It was a patch of unusually warm water that formed in 2013 in the Gulf of Alaska. The patch lingered and spread southward along the western coast, eventually stretching all the way to Mexico. By the summer of 2014, scientists were calling it the Blob.

Marine heat waves, times when temperatures at the surface of the ocean are much higher than usual, are nothing new. But climate change is making them more severe. The Blob was especially bad, breaking records for size, severity, and duration. At its peak, the Blob covered more than 1.5 million square miles (4 million sq. km), half the size of the United States excluding Alaska. In some areas, surface waters climbed 5°F to 9°F (2.7°C to 5°C) above normal. The Blob lingered for more than two years, into the summer of 2016. Did this heat wave kill common murres?

To find out, Piatt's team first looked at the fish murres eat. Did they move to cooler waters in response to the Blob? Maybe the fish moved north, south, farther offshore, or deeper in the ocean. Even if they had, murres can fly fast and far. They can dive deep. If the fish had relocated, Piatt's team concluded, the birds could have found them.

Next, the team looked at the ocean's tiniest inhabitants, phytoplankton and zooplankton. They found that in the years of the Blob, these organisms changed. Phytoplankton communities shrank in size. Zooplankton populations reshuffled. "The older, fatter, nutritionally richer zooplankton were replaced by . . . species that weren't as big and nutritionally rich," Piatt said.

The effects rippled to small ocean fish like capelins and juvenile pollock, which are food for common murres. As the water warmed, those prey fish became less abundant. Meanwhile, competition for those fish increased. In hotter water, big predator fish like cod, flounder, pollock, and hake eat more. Thanks to the Blob, predator fish were eating more of the small prey fish that common murres need.

Piatt's team concluded that common murres had been crushed in a climate vise. As the number of prey fish plummeted, competition for those fish increased. Murres couldn't survive the squeeze.

 ## COPING WITH CHANGE

Climate change doesn't affect just ocean birds. From coastlines to wetlands, grasslands to forests, climate change places birds in all kinds of habitats at risk. As Earth heats up, birds have to contend with more powerful storms. Droughts are increasing, worsening wildfires in many places. Around the world, spring weather is coming earlier, hot summers are lasting longer, and winters are growing more severe.

Each bird species is finely tuned to the environmental conditions in its habitat—conditions such as temperature, precipitation (rain and snow), and the timing of seasons. To survive warming global temperatures, birds must leave their homes and try to find suitable homes elsewhere. Scientists have noted that birds in the Northern Hemisphere are shifting their homes northward. (A comparable

PERIL ON THE HIGH SEAS

Seabirds are some of the most imperiled birds on the planet. Worldwide, almost half of all species are declining in number. What is behind the declines?

In 2019 an international team of scientists assessed the threats to seabirds worldwide and published their findings in the journal *Biological Conservation*. They identified the biggest problems as predators, entanglement in fishing gear, and climate change. First, cats and rats prowl for prey at many seabird colonies. These predators can take a big bite out of some populations. Second, many seabirds get accidentally hooked or tangled in commercial fishing nets and other gear. This "fisheries bycatch" is a serious threat to some species. Finally, climate change threatens to radically alter marine habitats. The changes to climate are happening fast, leaving birds with little time to adapt.

The report also flagged additional risks, such as plastic pollution in the ocean. Birds can die when they accidentally swallow plastic trash that floats in the ocean. Although plastic is not yet a significant threat to most seabird species, the threat is growing and could be bigger than scientists realize.

Another emerging threat is offshore wind farms. Many birds die when they collide with wind turbines near coastlines. Most environmental groups endorse offshore wind development because unlike burning fossil fuels, wind energy does not emit carbon dioxide and contribute to climate change. The way to make offshore wind farms bird-friendly, experts say, is to locate them in parts of the ocean not visited by at-risk birds.

Plastic trash is a growing threat to seabirds.

shift is happening in the Southern Hemisphere, with birds there moving southward.) They're also migrating and laying eggs earlier.

Scientists from the National Audubon Society calculated how 604 North American bird species could be affected as the climate continues to change. They used 140 million observations made by birders and scientists from 2000 to 2009 to map out each bird's range—where it lived in the first decade of the 2000s. They then used climate models to predict how each bird's range might shift in response to rising temperatures and changes in rainfall. The models showed that climate change would destroy bird habitats and put two-thirds of North American bird species at risk of extinction. The numbers showed that 100 percent of Arctic birds and 98 percent of birds in Canada's northern forest were vulnerable. Birds of other forests, waterbirds, grassland birds, and coastal birds were also at high risk.

Allen's hummingbird (*below*) is one example of an at-risk bird. The green-and-copper hummingbird breeds along the Pacific coast in a narrow strip from California to southern Oregon. It nests in open woodlands, wooded suburbs, and city parks. Scientists predict that climate change could make much of this strip unsuitable for the tiny bird. It could be restricted to just 32 percent of its breeding

range by 2050 and a mere 7 percent of its range by 2080. What happens then? The birds might relocate and find new habitat away from the coast. But they'll face competition in their new home. They'll have to compete with other hummingbirds for limited food and nesting sites. They may not survive.

Sea Change

TRIPLE THREAT

For birds that depend on the ocean, climate change presents a triple threat. First, as Earth warms, the oceans absorb excess heat. From 1900 to 2016, surface waters around the world warmed on average 1.3°F (0.7°C). This excess heat is altering ocean currents, shifting wind patterns, and causing marine heat waves. It's also making it harder for seabirds to find a meal. As the oceans heat up, small fish are moving farther offshore to cooler waters. That means seabirds have to fly farther from their colonies to find fish for themselves and their chicks—using up precious energy.

Second, as atmospheric carbon levels rise, excess carbon dioxide from the air dissolves into seawater. The extra carbon makes the water more acidic. This chemical change affects the shells of phytoplankton, zooplankton, oysters, clams, and corals. The shells don't form properly, and these organisms cannot survive without

As ocean temperatures rise, glaciers are melting. In turn, sea levels are rising.

their protective shells. As plankton and other prey populations decline due to acidification, the losses ripple through the ocean food chain, reaching prey fish and the birds that eat them.

Finally, rising temperatures melt glaciers and ice caps on land. This melting causes sea levels to rise, flooding coastlines. When coasts go underwater, birds of sea and shore lose safe places to nest. This is especially dangerous for birds that nest in low-lying areas, such as the black skimmers that nest on Atlantic islands along the coast of the Carolinas.

As ocean temperatures climb, scientists predict that marine heat waves like the one that caused the Blob will become increasingly common. A 2018 *Nature* report warned that if global warming continues unabated, ocean heat waves will become forty-one times more frequent by the end of the twenty-first century. Those heat waves will likely grow hotter and last longer. The changes, the researchers argue, could push "marine organisms and ecosystems to the limits of their resilience."

The heat wave that killed common murres also reached other seabirds. Puffins, auklets, kittiwakes, fork-tailed storm petrels, fulmars, and shearwaters also died. Hundreds of sea lions starved along the California shore. Dead fin whales and sea otters washed up onto the Alaskan coast.

The Blob finally subsided in 2016. But for common murres, the effects will linger. Because most of the dead birds were adults of breeding age, the breeding population took a big hit. What's more, biologists who monitor murre colonies documented twenty-two separate breeding failures—when a colony produces zero chicks— during and after the die-off. Scientists strongly suspect that hunger was to blame, because seabirds will skip breeding if they don't have enough to eat. Piatt concludes that one breeding failure in a year would normally be uncommon. "To have repeated failures at large, important colonies? That's unprecedented."

BRINGING BACK NATURE

We must abandon our age-old notion that humans and nature cannot mix, that humans are here and nature is somewhere else. Starting now, we must learn how to coexist.

—American entomologist Douglas Tallamy

I n 2000 Doug Tallamy and his wife, Cindy, moved to 10 acres (4.1 ha) of former farmland in southeastern Pennsylvania, near the Maryland border. They wanted a home in the country, a slice of nature close to his job as an entomology professor at the University of Delaware. But after moving in, they realized that their new property was a mess. More than a third of the land was overrun with thorny, weedy plants. The weedy plants were not native to North America but instead came originally from Europe and Asia. Earlier generations of Americans had imported the plants from overseas to grow in their gardens. The weedy invaders were smothering and squeezing out native trees, shrubs, and wildflowers, taking up space, light, and water that the native plants needed.

Many people invite birds into their yards with nesting boxes.

The Tallamy family made it their goal to restore the native ecosystem on their property. In their free time, they would hack back the invaders and encourage the native plants, the ones that had evolved to live in the woods of southeastern Pennsylvania.

Armed with a hand saw, clippers, heavy gloves, and herbicides, the family got to work. But as Doug Tallamy slashed and sweated his way through the wild tangle, he noticed a curious pattern. The native plants—the pin oaks and red maples, the black cherries and black walnuts—all showed signs of being eaten by insects. Their leaves were dotted with neatly chewed holes or decorated with delicately nibbled edges, clear clues that insects had dined. But the invaders—the Norway maples and autumn olives and Japanese honeysuckles—showed little or no damage. Insects were eating the native plants but avoiding the non-native ones. Why do insects eat some plants and not others? The answer involves evolution.

THE CONQUERORS

Consider caterpillars (the larvae of moths, butterflies, and sawflies). Nearly all caterpillars eat plants. But caterpillars can eat and digest only a few plants—known as their host plants—that they have lived alongside for thousands of generations.

Caterpillars are voracious eaters. To keep themselves from being eaten by caterpillars and other insects, plants produce nasty chemicals. These chemicals make a leaf toxic or bad-tasting to repel insects. This defense keeps the plants from being chewed to the ground. But over time, insects can evolve and adapt to these chemicals.

A good example is monarch caterpillars, whose host plants are milkweeds. Even though milkweed leaves contain toxic chemicals, through evolution, monarch caterpillars developed the ability to ingest these chemicals without getting sick. They sequester (isolate) the toxins in their bodies, where the substances cannot hurt them. So monarch caterpillars thrive on milkweed leaves, but other kinds of leaves are poisonous to them.

In one portion of the food chain, monarch caterpillars eat milkweed plants and birds feed on monarch caterpillars.

Like monarchs, most caterpillars won't eat just any plant because they can't eat just any plant. They can digest or detoxify (make less toxic) only the chemicals from their host plants, which they have repeatedly encountered over a long time—sometimes for thousands of years.

Yet many plants growing in North America, like many of those in the Tallamy garden, have not lived on the continent for thousands of years. In just the past few hundred years, people brought these non-native plants from other parts of the world, either accidentally or deliberately. Sometimes seeds of weedy plants, such as Russian knapweed, creeping thistle, and cheatgrass, hitchhiked to the continent inside hay bales, in bags of grains, or stuck onto the fur of livestock. Plants like Norway maple, Japanese honeysuckle, and kudzu came when North Americans planted them for timber, for food, or for their good looks. At one time, the US government paid explorers to travel to distant lands and seek useful plants. That's how the Bradford pear got to the United States. The US Department of Agriculture introduced two hundred thousand different types of plants from around the world.

When a plant is introduced to new territory, it lives out of range of most of its natural enemies. Most of the diseases and insect pests that might kill it are not found in its new home. Plant sellers advertise this feature as an advantage: *Look, gardeners, this plant has no insect pests!* But without insects eating it, a non-native plant can outcompete and overgrow native plants. Some non-native plants are so aggressive that they become invasive species, harming the environment or causing headaches for people. They may invade forests, take over meadows, or choke waterways. They may pull down electric lines or yank down tree limbs, the way thick, ropy kudzu vines do in the US Southeast. They may even fuel the spread of wildfires. For instance, in the American West, cheatgrass grows densely across the ground. It is highly flammable, so when wildfires break out, burning cheatgrass makes them more severe.

ATTACK AND COUNTERATTACK

A plant's life may look peaceful, but look closer. Plants are constantly doing battle with the insects that want to eat them. Worldwide, plant-eating insects make up 37 percent of all species. And those insects are hungry.

Plants can't run away from hungry insects, so what's a plant to do? They fight back with an arsenal of weapons. Many plants cover themselves with dense hairs, waxy layers, or thick bark to avoid insect jaws. But the most potent weapon in the plant world's arsenal is poison. Plants pump their leaves full of toxic chemicals to keep insects from chewing them. These chemicals may taste bad or be downright poisonous to insects that try to dine.

But over time, insects might develop the ability to tolerate a certain plant's chemical. They might produce a substance that allows them to digest or detoxify the chemical. With this new ability, the insect gains a big survival advantage: a food source that its competitors can't eat, but it can. But over time, the plant can fight back by making a more powerful poison. In response, the insect evolves greater tolerance—and on and on. This back-and-forth can lead to an arms race, where plants evolve ever-stronger defenses and insects evolve to keep up.

This is called coevolution, when one species evolves in response to the other. But when non-native plants arrive, they can throw a carefully balanced ecosystem out of whack. Non-natives grow without the insect enemies that normally keep their growth in check. So they can grow out of control and throw a once-balanced ecosystem into disarray.

NO CATERPILLARS, NO CHICKADEES

As Doug Tallamy hacked away at all the invasive species in his overgrown yard, a light bulb clicked on. He already knew that the invasives had taken over largely because caterpillars and other insects wouldn't touch them. Not being eaten gives invasive plants a big advantage. But Tallamy realized something else. If our yards are full of non-native plants, then our native insects must have nothing to eat. And since birds eat insects, could that mean our birds have less to eat too?

As Tallamy knew, different birds eat different foods. Sparrows snack on seeds, catbirds gobble berries, robins nibble worms, hummingbirds drink nectar. But those are all foods *adult* birds eat. Baby birds, with their mouths wide open, waiting in the nest to be fed by a parent, don't usually eat seeds or berries or nectar. They eat bugs. And most bird parents load up their offspring with one particular type of bug: caterpillars. Caterpillars are to baby birds what smooshed peas are to baby humans—soft, packed with nutrition, and easy to gulp down.

Birds need caterpillars, and caterpillars need native plants. When Tallamy compared the native and non-native plants in his yard, he saw that the natives supported thirty-five times as many caterpillars as the non-natives. And this caterpillar abundance peaked in the spring, right when many birds were breeding. Tallamy knew what this meant: yards full of non-native plants may look pretty, but they offer much less food for birds. "No wonder our birds are struggling," he wrote in his best-selling book *Bringing Nature Home*.

Tallamy has conducted many studies of native plants, insects, and birds in food chains. According to one of his studies, a single brood of five baby chickadees can gulp down more than nine thousand caterpillars in the sixteen to eighteen days before they're ready to leave the nest. His findings show that one of the most powerful ways we can restore habitat for birds is by encouraging insects to live there, especially caterpillars. And encouraging caterpillars starts with native plants.

One of Tallamy's studies examined caterpillars on native and non-native plants in eight eastern states: Virginia, Maryland, Delaware, Pennsylvania, New York, New Jersey, Connecticut, and Rhode Island. From those findings, he created a native plant index, which ranks different types of plants by the abundance of caterpillars they support. First place on the index went to oak trees, which supported 534 species of caterpillars. Tied for second place were cherries and plums, which supported 456. Willows came next with 455.

The study confirmed his suspicion that gardeners can play a pivotal role in building habitat for birds. Tallamy has written a second book, *Nature's Best Hope*, in which he argues that growing native plants is a way for ordinary people everywhere to help birds and other animals—by turning our backyards into a network of restored habitat that helps struggling wildlife populations.

 ## URBAN BIRDS

Desirée Narango, an urban ecologist at the University of Massachusetts, studies the interaction of living things and their environment in cities. "I grew up in Baltimore, Maryland," she says. "The green space I had access to was my backyard and a park." She studied ecology in college and after graduation did fieldwork on birds in South America and Central America. But she found herself drawn to questions that hit closer to home: Which birds are able to live in cities and why? Why do birds use some urban spaces and avoid others? What makes an urban green space into a good or bad habitat for birds?

Narango focused her questions on Washington, DC, where a

citizen science project called Neighborhood Nestwatch was already underway. Neighborhood Nestwatch uses the Washington metropolitan area as a living laboratory. Participants invite scientists to come into their backyards and schoolyards. The scientists capture backyard birds like catbirds and chickadees in mist nets. They tag birds with colored bands on their legs and release them. Throughout the year, participants keep track of any tagged birds they see, identifying them by band color, and report the results back to the scientists. The program is helping researchers study the status and trends of the backyard bird population in the city.

Narango tapped into Neighborhood Nestwatch to investigate backyard food chains. Just as John Piatt had done along the Pacific coast and Christy Morrissey had done in the ponds that pothole the Canadian prairie, Narango worked painstakingly, step by step, through the food chain. She began by assessing the plants growing in each participant's yard. "You'd go into a neighborhood that would look really nice," she explains. "But then when you start identifying the species, these are all Norway maples and Japanese cherries"—non-natives. Other neighborhoods would be filled with natives, such as white oaks and flowering dogwoods.

Next, she searched the different plants for caterpillars (which she calls "big, delicious items of food [for birds] that have great protein") and spiders—"smaller but also high [in] protein". By collecting and weighing all the caterpillars and spiders she found, she could estimate the number of insects in each yard. On the introduced plants, she often found one caterpillar or none. But in native trees such as oak trees, she found dozens of them.

Then she turned her attention to Carolina chickadees, common backyard birds in cities and suburbs of the southeastern United States. She set up mist nets in yards and played a recording of the chickadee's familiar call: *chikadeedeedeedee*. "Chickadees are insanely easy to catch," she said. "When we play those chickadee calls, basically everyone comes in to investigate."

She banded the captured chickadees and released them. Then she returned throughout the year, slinging a pair of binoculars and following banded chickadees from yard to yard to see which yards they were visiting. Their activities were revealing. One chickadee pair visited a Sunoco gas station, where native oak trees had been planted years earlier. "If the right plants are there, [birds] will find them," she observed.

By following the chickadees from year to year—and by training video cameras on the nests to learn what kinds of foods the chickadees were bringing their young—she was able to make connections between the types of plants in a yard and the overall health of chickadee populations. She could ask, Did the male chickadee get a mate? If so, how many eggs did the female lay? How many of the young fledged, or left the nest? How many of those fledglings survived?

Gathering all those threads of data, she wove together a picture of the chickadee's urban habitat. The picture revealed whether each yard was supporting population growth or leading to population decline. She could see whether each yard was good or bad habitat for chickadees.

Together with Tallamy and Peter Marra of the Smithsonian Migratory Bird Center, who helped with the project, Narango published the findings in science journals. The studies revealed that chickadees were more likely to nest in yards full of native plants. Native oaks, cherries, and maples made good chickadee habitat because they supported more caterpillars. And the studies showed that the amount of native plants in a yard was critical. Once the mass of native plants fell below 30 percent of all vegetation, chickadee populations plummeted. The birds couldn't find enough food to thrive.

LAWN GONE

Americans love their lawns. In the United States, more than 45 million acres (18.2 million ha) are carpeted with lawns—an area the size of North Dakota. And we add 320,000 acres (130,000 ha) of lawn every year.

But lawns are not natural ecosystems. They are planted with non-native grasses. Although lawns may look lush and inviting, they don't provide good habitat for birds.

Maintaining all that lawn is a huge undertaking. The average homeowner devotes a full workweek every year to mowing. And US lawn care accounts for 70 million pounds (32 million kg) of insecticides and herbicides every year—ten times the amount used for farming. All that chemical-laced runoff seeps into groundwater and threatens wildlife and human health.

Many bird lovers are kicking the lawn habit. They are digging up parts of their lawns and replacing them with native trees, shrubs, and wildflowers. Doug Tallamy suggests starting small and working gradually. Don't rip out the lawn until you have a plan for the area, he says. And he stresses, "This can be a hobby. You don't have to do it all at once."

"HOMEGROWN NATIONAL PARK"

Narango wants people to stop thinking of their yards as "pretty things that are separate from nature." She continues, "If we can get everybody that has a yard to start planting native plants, then that's a huge amount of green space and a huge amount of habitat to help support our declining bird populations."

This shift is already underway, as more and more gardeners plant native vegetation and invite insects into their yards. "We're shifting from, 'Eww, bugs are icky. We don't want them on any of our plants,' to understanding that these things are all connected and we can't have songbirds unless we have these healthy food webs [food chains]," she says.

Backyards and schoolyards, parks and commercial green spaces all give people an opportunity to restore habitat for birds. It's an important point that Narango and Tallamy are trying to teach people. Birds need habitat not just far away, in a wilderness preserve, the ocean, or a distant national park. They need habitat here in the places where we live, work, and play. And we can build this habitat with native plants.

Narango's work in Washington, DC, shows the power of doing this, especially when people band together. She found that some neighborhoods supported more birds than others, largely because people had planted different types of trees there. "Small choices in a neighborhood . . . can really make a meaningful improvement for these birds," she says.

A black-capped chickadee feeds on a caterpillar.

In *Nature's Best Hope*, Tallamy writes that ordinary backyards represent a powerful opportunity to restore a great deal of lost habitat all over the continent. If Americans were to convert just half of their lawns to bird habitat, the restored habitat would amount to 20 million acres (8.1 million ha). As Tallamy points out, that's "bigger than the combined areas of the Everglades, Yellowstone, Grand Teton, Canyonlands, Mount Rainier, North Cascades, Badlands, Olympic, Sequoia, Grand Canyon, Denali, and the Great Smoky Mountains National Parks." He even has a name for this park created where we live, work, and play: "We call it Homegrown National Park."

9

HOW YOU CAN HELP

We have taken over the earth and the sea and the sky, but with skill and care and knowledge, we can ensure that there is still a place on Earth for birds in all their beauty and variety—if we want to—and surely, we should.

—British broadcaster and naturalist
Sir David Attenborough

On a cold winter morning, I awoke to a noisy chorus of birds—an unexpected sound for the snowy days of February. Bleary-eyed, I headed downstairs to see what was going on. My cat Cleo was already at the living room window. She sat in a crouched position, her tail flicking. I stepped up to the window, looked outside, and found that my snow-covered backyard was full of birds.

Most of them were robins, which can form big, chattering flocks in the winter months. A flock of sixty or more had dropped in on my yard, looking for food. The robins were joined by a few

A robin mother feeds worms to chicks in the nest.

cedar waxwings, sleek birds that look as if they wear black masks and have had their tails dipped in golden ink. The birds had gathered in the trees that ring my yard. Some were hopping on the snow-covered grass. They were taking turns flying to a shrub just outside the living room window, a native viburnum bowed down with crimson berries. A wave of birds would land on the branches, gobble a batch of berries, and return to the distant trees. Then the next wave of birds would swoop in for their turn at the berry buffet.

I watched them for about an hour and then had to get on with my day. When I returned to the window hours later, I saw that the flock had moved on. The yard was quiet again. There were just our usual winter residents—some sparrows, a few juncos, and a pair of cardinals. The only sign that the noisy flock had been there were the few bright berries they had left scattered on the snow.

With displays such as this, sometimes it can be hard to believe that birds are really in trouble. Yet decades of monitoring reveal a bird world in deep distress.

How You Can Help

Birds need our help. The loss of three billion birds is a sign of nature in crisis. As you think about this crisis, remember what happened to the passenger pigeon. Remember what almost happened to the peregrine falcon and the bald eagle. Ask what you can do to protect birds and their habitats so they will continue to live with us.

We can all do our part. "The important thing is to take action," says Gary Langham, then chief scientist at the Audubon Society. "There's something everyone can do in their lives and in their communities to make it a better place for birds and people." You can help birds in some simple and fun ways.

 ## GET TO KNOW BIRDS

There's a reason bird-watching is popular. Birds are fascinating animals. You don't need fancy equipment. You just need your eyes, ears, and curiosity. You don't even have to go far. You can watch birds right outside your home or at a nearby park or nature preserve. Even small yards and parks can host migratory and nonmigratory birds. Some migratory birds will linger for the season. Others will touch down briefly for stopovers on their long journeys.

If you have a smartphone, consider downloading the Cornell Lab of Ornithology's Merlin Bird ID mobile app. It will help you identify the birds you see. You can also buy or check out from the library a basic field guide for beginners, such as *Stokes Beginner's Guide to Birds*. You'll find one Stokes guide for birds of the western United States and another for the eastern states. The birds are grouped by color and then arranged within the group from smallest to largest. That makes it easy to quickly identify that bird you see flitting through the branches.

If you are interested in raptors, you can go to a hawk watch. Hawk-watching sites are found all over North America. Visit the

BUSTING BIRDER STEREOTYPES

Black Birders Week launched in 2020, sparked by a racist incident in New York City. In May 2020, Black bird-watcher Christian Cooper asked a white woman to put her dog on a leash as required by Central Park rules. She refused and called 911, falsely claiming to authorities that Cooper was threatening her and pointing out his race as though it made him dangerous. A smartphone video of the incident spread on social media, sparking outrage and drawing attention to the racism that people of color face in outdoor spaces.

People of color have traditionally been excluded from outdoor activities, including bird-watching. A 2011 study by the US Fish and Wildlife Service found that 93 percent of birders are white and that 24 percent of white people participate in bird-watching. Only 7 percent of Black people are birders. When Black people take part in bird-watching events, white birders sometimes treat them as outsiders or unwelcome. Black birders might even be harassed, as the incident in Central Park showed.

In response to the incident, an online group of Black scientists launched Black Birders Week, a series of events to highlight Black scientists and bird-watchers. Event organizer Corina Newsome, a wildlife biologist who studies seaside sparrows, says the goal is to get more people hooked on birds while making outdoor spaces welcoming to people of color.

"It's not only good for the people that have historically been excluded, it's good for the field [study of birds] as well," she says. She points out that diverse perspectives are needed to solve problems in wildlife conservation. "If everyone has had the same life experiences, you're less likely to have the solution to a problem."

Keith Russell from Audubon Pennsylvania conducts a breeding bird census in a Philadelphia park.

website of HawkWatch International, and click on a US state or Canadian province to find one near you.

 ## MAKE WINDOWS SAFER

More than a million North American birds die every day from collisions with windows. But you can make your windows safer for birds. If you have a dangerous window—such as a big picture window—you might want to buy or make simple window treatments. They break up the reflections in a window so that birds won't fly into it. Check out www.birdsavers.com/make-your -own for instructions on how to make these bird savers. Be sure to check with an adult before installing anything new on your home windows.

The cords on this window break up reflections, helping prevent bird strikes.

FEED BIRDS SAFELY

Before putting up bird feeders, make sure the area is safe. You don't want to attract birds if cats are roaming or the birds could crash into a window. If you do put up a feeder, you will need to locate it very close to the window—within 3 feet (0.9 m)—so that birds won't be traveling very fast if they do fly into the glass. Or you can keep feeders as far away from your windows as possible—at least 30 feet (9.1 m).

Fill your feeder with high-quality bird food. Different birds eat different kinds of food. To attract woodpeckers, nuthatches, and titmice, try a mixture of peanuts, other nuts, and dried fruit. Black-oil sunflower seeds are a good choice because they will attract many kinds of birds. Cardinals, blue jays, chickadees, nuthatches, and sparrows love them. Offer different seeds in different containers.

TURN DOWN THE LIGHTS

You can join the movement to turn down lights that can disorient migrating birds. Use outdoor lighting only when and where you need it. Make sure outdoor lights are pointed down instead of up into the sky. Close curtains, blinds, and shades at night to block interior light to keep night-migrating birds from flying into windows.

KEEP CATS INDOORS

If you're a cat owner, you can protect birds and your cat. Remember that it's healthier for your cat to stay indoors. You could turn an enclosed porch or patio into a "catio," where your kitty can enjoy fresh air and sunshine but won't be able to attack birds. You can even train your cat to walk on a cat-friendly leash when you're both outdoors. This too will give your pet sun and fresh air but not the

How You Can Help

freedom to hunt birds. (It's easier to get a kitten used to a leash than a grown cat.)

People who don't want to keep cats indoors can still help curb the killing. Putting a bell on your cat's collar can alert prey to its presence. The ringing bell will warn birds, giving them time to fly off. Some places even sell kitty clothes designed to make your cat less lethal to birds. One product is a brightly colored cover that slips over your cat's collar. Since songbirds see bright colors especially well, they will notice the colored collar and fly off before your cat has a chance to pounce. Another product, the colorful CatBib, gives a visual warning to birds and interferes with a cat's paws when it tries to pounce. You can also make your cat less deadly by trimming its claws regularly. Finally, you can keep your cat indoors a little more than you do now, to limit the damage it can do. And you could decide that your next cat will be an entirely indoor animal.

Whatever you do, never abandon a pet cat. If you are no longer able to care for your pet, surrender the cat to a local animal shelter to give it a chance of being adopted. And always check with an adult before making any pet care decisions.

 ## SUPPORT SUSTAINABLE AGRICULTURE

You can support bird-friendly agriculture in your area by buying local, organically grown food. Farmers markets are great places to buy directly from local growers. At market stalls, look for signs saying that the fruits and vegetables on sale there were grown using organic farming practices. Many grocery stores also sell organic produce. By supporting farmers who grow this produce, you are supporting birds. Christy Morrissey points out that buyers decide how food is grown by "voting" with their food-buying dollars. "We eat the food," she says. "We have a lot of control."

FIGHT CLIMATE CHANGE

Climate change is hurting birds. This problem cannot be solved by one person acting alone, but you can help in many ways. You can make changes that reduce your own carbon emissions, such as lowering your home's thermostat in winter, thereby burning less fossil fuel. You can share what you're doing to fight climate change with others, either in person or on social media. You can join in political action by contacting congressional representatives and other lawmakers, urging them to pass laws to reduce carbon emissions.

One of the most rewarding ways to fight climate change and help birds is to plant a tree. All plants remove carbon dioxide from the air during photosynthesis and store it in their wood, roots, and shoots. As giants of the plant world, trees can store a lot of carbon dioxide.

Planting trees is a win-win for the environment. Trees absorb carbon dioxide from the atmosphere and serve as homes for birds and other animals.

In addition to fighting climate change, trees provide habitat for birds. Doug Tallamy has researched which native trees support the highest numbers of caterpillars, providing more food for baby birds. You can look up the best trees for your area on the National Wildlife Federation (NWF) website. Or talk to tree experts at a nearby botanical garden to learn which native trees are best for your yard.

Before you plant a tree, check with an adult and be sure to consider the tree's final size. One thing you can count on: the tree will get bigger. And some trees will get a lot bigger! But even if your yard is small, you can probably find a small tree that fits perfectly.

 ## INVITE BIRDS INTO YOUR YARD

By growing native trees and plants, you can turn your yard into a rest stop for migrating birds or a safe home for birds that are nesting or spending the winter. You can also help build bird habitats in your community. Bird lovers are creating bird-friendly gardens in places all across North America. They may need volunteers to help plant or maintain these gardens. Do a Google search to find bird-friendly gardens in local parks, at schools, and in your neighborhood.

You'll want to fill your bird habitat with native plants. Remember: No native trees, no caterpillars. No caterpillars, no chickadees. Check out the Audubon and the NWF websites to find plants native to your area.

Swear off pesticides in your habitat patch. Pesticides can be toxic to birds and can kill the insects that birds need to survive. When you buy plants for your garden, ask if the plants were treated with pesticides. If so, shop elsewhere. Seek garden suppliers that sell organic or pesticide-free plants.

Add other features that birds need. Create a brush pile by placing large branches loosely on the ground and layering more branches on top. Be sure to leave open pockets between the layers. This gives songbirds a place to hide from predators. Set up a birdhouse or nesting box on a post or tree. It will give a home to birds that normally nest in tree cavities and other holes. Buy a nesting box or build one yourself. Visit the NestWatch website at nestwatch.org to learn about nesting requirements for different birds and to download construction plans. Add a birdbath or another source of water, giving birds a place to come for a drink or a bath. You don't have to buy a birdbath. Make one by filling a saucer or pie plate with water and placing it on a stump, rock, or upturned flowerpot. Once you've made your habitat patch, sit back and watch the birds show up!

 ## SHARE WHAT YOU SEE

Scientists who monitor and study birds need citizen scientists— people like you who report the birds they're seeing in their yards, neighborhoods, and wild places. Your participation matters. It connects you to the beauty of birds and nature while helping scientists monitor bird populations. Scientists use information provided by citizen scientists to find out which birds are in trouble and where they are. Participation in citizen science projects typically doesn't cost anything.

Join the oldest citizen science project on the continent, the Christmas Bird Count, run by the National Audubon Society. Sign up in November through the Audubon website. Counts are held from December 14 to January 5 every year, and your local count will be on one of those days. You and other volunteers will report to a designated circle that is 15 miles (24 km) wide. Within your circle, the volunteers will break up into small groups. If you're a beginning birder, you'll be teamed up with at least one experienced birder. Your group will follow an assigned route, and together you'll count and report every bird you see.

How You Can Help

COUNTING BIRDS

The Christmas Bird Count is the longest-running citizen science project in America. It began more than a hundred years ago. In the 1800s, many people participated in a popular holiday tradition called Christmas side hunts. Friends and neighbors divided into teams, went into the countryside, and competed to kill the most birds.

In 1900 US ornithologist Frank Chapman proposed a new tradition—counting birds, rather than killing them. That first year, twenty-seven observers counted around 90 species of birds in twenty-five locations across the United States and Canada. Since then the counts have been held every year and have grown in popularity. In 2019—the most recent count as of this writing—more than eighty-one thousand observers counted 2,566 species of birds all over the Western Hemisphere.

Carole Hirsh and Clive Pinnock, manager of the Okeeheelee Nature Center in West Palm Beach, Florida, count birds at the Christmas Bird Count.

You can also participate in the Great Backyard Bird Count over four days in February. The program is run by the Cornell Lab of Ornithology, the National Audubon Society, and Birds Canada. You'll join hundreds of thousands of bird counters around the world. You can count the birds you see for as little as fifteen minutes, identifying the birds with the Cornell Lab of Ornithology's free eBird mobile app. You can submit your sightings through the app or through the eBird website.

Another great project if you want to count birds from home is Project FeederWatch, which runs from mid-November to mid-April. This project is run jointly by the Cornell Lab of Ornithology and Birds Canada. To participate, you choose a part of your yard that is easy to monitor. A good area is where you can easily see a bird feeder, but even if you don't have a feeder, you can still count the birds. For two days in a row, you'll count all the birds you see: birds that visit a feeder or birdbath, birds that visit trees and shrubs, or hawks and other predators attracted by the birds at your feeders. Even counting one time each winter will give scientists valuable data.

* * *

Birds connect us to the natural world. You can glance at the cardinal that flashes past your window or listen to a sweet chorus of birds on a spring morning or watch a hawk soaring overhead. Then you can go back to your normal life. That's the easy thing to do. But it won't help birds.

To give birds a hand, start with one of the steps in this chapter. Go to a hawk watch. Join a citizen science project. Plant a native tree. Then try another step. And another. You don't have to do everything at once. The truth is, even small steps can help birds.

The disappearance of birds is a big problem, but it's one we can fix. We have the power to make the world a safe and healthy place for birds—and for all the other creatures that share our planet. Let's get started.

When I heard the news in 2019 that nearly three billion North American birds had disappeared, I was dumbstruck. I searched my memory. Had I seen fewer birds lately? Maybe. It was hard to know. As I pondered what the loss of almost three billion birds meant, my thoughts turned to one particular bird: the bald eagle.

Growing up in Pennsylvania in the 1970s and 1980s, I never saw a bald eagle. Not one. I was taught that the birds were endangered and had disappeared from my part of the country. But a few years ago, I saw a pair of bald eagles fly in wide, sweeping circles over my suburban neighborhood. I stood outside watching them, my neck craned back. My neighbor, a retired wildlife biologist in his eighties, joined me. For a few moments, we shared in the joy of watching eagles soar over our houses.

Bald eagles, once long gone from Pennsylvania, are again plentiful. The bald eagle represents what all birds can and should be—abundant, valued, and thriving. After writing this book, I believe we can reach that goal, but only if we work together to make the world a safe place for birds.

This book would not have been possible without the generous assistance of a number of scientists and experts who took time out of their busy schedules to talk or meet with me, share their experiences, and answer my many questions. In particular, I would like to thank David Barber of Hawk Mountain Sanctuary, Margaret Eng of Environment and Climate Change Canada, Daniel Klem Jr. and Peter Saenger of the Acopian Center for Ornithology at Muhlenberg College, Michael Mesure of FLAP Canada, Christy Morrissey of the University of Saskatchewan, Desirée Narango of the University of Massachusetts—Amherst, and Keith Russell of Audubon Pennsylvania. I also appreciate the many contributions to the study of bird declines made by Peter Marra of the Smithsonian Migratory Bird Center and Doug Tallamy of the University of Delaware.

Finally, I wish to thank my readers. The future of our planet rests in your hands.

Glossary

biodiversity: the variety of different plant and animal species in a place or region

breeding ground: the place where an animal mates

carnivore: an animal that eats meat

citizen science: the study of the natural world by nonscientists in collaboration with scientists

climate change: the warming of Earth due to increased levels of carbon dioxide in the atmosphere, caused by the burning of fossil fuels

conservation: careful preservation, protection, or restoration of natural resources, habitat, vegetation, or wildlife

control group: subjects in an experiment that are not given a procedure or substance so that they might be compared to subjects that are given the procedure or substance

ecologist: a biologist who studies the relationship between living things and their environment

ecosystem: a biological community of living things and the environment in which they live

ecotoxicologist: a scientist who studies the effects of toxic substances, especially pollutants, on wildlife and the environment

endangered species: a species that is at serious risk of extinction

entomologist: a biologist who specializes in studying insects

evolve: to change over time, often in a way that enhances an organism's chances of survival and reproduction. Over many generations, this change can lead to the development of different species.

extinct: having no living members. A species becomes extinct when the last individual of the species dies.

fledge: to leave the nest after developing feathers necessary for flight

flyway: a flight path used by large numbers of birds while migrating

food chain: the flow of food energy through the environment, often starting with plants or plankton, which are eaten by small animals, which are then eaten by larger animals, which might be eaten by other animals. Since most animals eat a variety of different kinds of foods, foods chains can be complex.

fungicide: a chemical used to kill fungi

habitat: the natural home of a living thing, such as a prairie, forest, or lake. A habitat provides plants and animals with food, shelter, protection, and partners for mating.

herbicide: a chemical used to kill plants

herbivore: an animal that eats plants

insecticide: a chemical used to kill insects

instinct: inherited knowledge that directs the behavior of animals, such as for nest-building, mating, and migration

invasive species: plants and animals that are not native to an ecosystem and whose introduction causes harm. Invasive species might prey on, compete with, or spread diseases to native species.

light pollution: artificial lighting that can interfere with animal behavior by changing the appearance of the night sky

mass extinction: a widespread extinction of living things during a relatively short period. Earth has experienced five mass extinctions. Many scientists believe human activities are causing a sixth extinction in modern times.

migrate: to move seasonally from one region or climate to another for survival and breeding

native: living or growing naturally in a particular place

nectar: the sweet liquid produced by flowers

nest: to prepare a place to raise young

ornithologist: a biologist who specializes in the study of birds

pesticide: a chemical substance for killing insects or other living things that are considered pests

photosynthesis: when plants and some other organisms use sunlight, carbon dioxide, and water to make food

pollinator: an insect or other animal that transfers pollen within or between flowers, helping them make seeds

predator: an animal that kills and eats other animals

prey: an animal that is hunted or killed by another animal for food

range: the region in which a plant or animal naturally lives

raptor: a meat-eating bird, such as a hawk, eagle, owl, or vulture. Also known as birds of prey, raptors have hooked beaks and large sharp talons.

species: a group of living things of the same type. Male and female members of the same species can breed (produce offspring) with one another.

wintering grounds: the place where a migratory animal spends the winter

Source Notes

4   Barry Yeoman, "Why the Passenger Pigeon Went Extinct,"
    Audubon, May/June 2014, https://www.audubon.org/magazine/may
    -june-2014/why-passenger-pigeon-went-extinct%0A.

5   Joel Greenberg, *A Feathered River across the Sky: The Passenger
    Pigeon's Flight to Extinction* (New York: Bloomsbury, 2014), 15.

6   Greenberg, 53.

6–7 Yeoman, "Why the Passenger Pigeon Went Extinct."

7   Yeoman.

12 Peter Marra and Chris Santella, *Cat Wars: The Devastating
    Consequences of a Cuddly Killer* (Princeton, NJ: Princeton
    University Press, 2016), 49.

16 Carl Zimmer, "Birds Are Vanishing from North America," *New
    York Times*, September 19, 2019, https://www.nytimes.com
    /2019/09/19/science/bird-populations-america-canada.html
    ?searchResultPosition=2.

16 Zimmer.

17 Karin Brulliard, "North America Has Lost 3 Billion Birds in
    50 Years," *Washington Post*, September 19, 2019, https://www
    .washingtonpost.com/science/2019/09/19/north-america-has-lost
    -billion-birds-years/.

17 Ed Yong, "The Quiet Disappearance of Birds in North America,"
    Atlantic, September 19, 2019, https://www.theatlantic.com/science
    /archive/2019/09/america-has-lost-quarter-its-birds-fifty-years
    /598318/.

17 Brulliard, "North America Has Lost 3 Billion Birds."

21 Yong, "The Quiet Disappearance."

21 "New Study Finds US and Canada Have Lost More Than One in
    Four Birds in the Past 50 Years," BringBirdsBack, accessed August
    16, 2021, https://www.3billionbirds.org/press-release.

22 "E. O. Wilson Receives Humane Society of New York's Humane
    Medal," E. O. Wilson Biodiversity Foundation, January 23, 2017,
    https://eowilsonfoundation.org/e-o-wilson-receives-humane
    -society-of-new-yorks-humane-medal/.

22 Daniel Klem, interview with the author, September 23, 2020.

23 Klem.

24 Klem.

24 Klem.

24 Klem.

25 Klem.

25 Klem.

28   Keith Russell, phone interview with the author, November 4, 2020.

31   CNN Wire Service, "1,500 Birds May Have Flown into Philadelphia Skyscrapers in One Day," *San Jose (CA) Mercury News*, October 9, 2020, https://www.mercurynews.com/2020/10/09/1500-Birds -may-have-flown-into-philadelphia-skyscrapers-in-one-day/.

31   Frank Kummer, "Up to 1,500 Birds Flew into Some of Philly's Tallest Skyscrapers One Day Last Week. The Slaughter Shook Bird-Watchers," *Philadelphia Inquirer*, updated October 9, 2020, https://www.inquirer.com/news/birds-center-city-philadelphia -audubon-october-2-2020-20201007.html.

32   CNN Wire Service, "1,500 Birds."

32   Russell, interview with the author.

32   Russell.

32–33 Russell.

33   Kummer, "Up to 1,500 Birds."

33   Russell, interview.

35   Michael Mesure, discussion with the author, March 19, 2021.

35   Mesure.

36   Mesure.

37   Russell, interview.

38   Marra and Santella, *Cat Wars*, 166.

41   Natalie Angier, "That Cuddly Kitty Is Deadlier Than You Think," *New York Times*, January 30, 2013, https://www.nytimes.com /2013/01/30/science/that-cuddly-kitty-of-yours-is-a-killer.html.

42   Angier.

44   "Paradise for Some—but an Ongoing Extinction Crisis for Birds," American Bird Conservancy, accessed August 16, 2021, https:// abcbirds.org/program/hawaii/.

45   Mark Price, "'Massacre' of Endangered Birds Linked to Hawaii's Booming Population of Invasive Cats," *Sacramento Bee*, September 7, 2020, https://www.sacbee.com/news/nation -world/national/article245354315.html.

47   Marra and Santella, *Cat Wars*, 164.

47   Marra and Santella, 172.

48   Lida Maxwell, "Another Silent Spring?," *Los Angeles Review of Books*, May 3, 2020, https://lareviewofbooks.org/article/another -silent-spring/.

52   Keith L. Bildstein, "A Brief History of Raptor Conservation in North America," in *The State of North America's Birds of Prey*, eds. K. L. Bildstein, J. Smith, and E. Ruelas (Orwigsburg, PA: Hawk Mountain Sanctuary, 2008), 14.

53   Rachel Carson, "Road of the Hawks," in *Lost Woods: The Discovered Writing of Rachel Carson* (Boston: Beacon, 2010), 32.

55   Rachel Carson, *Silent Spring* (New York: Houghton Mifflin, 1962), 2–3.

58   Emily Dickinson, "The Bobolink Is Gone—," H'llo Poetry, accessed February 23, 2021, https://hellopoetry.com/poem/2455/the-bobolink-is-gone/.

59   Elizabeth Royte, "The Same Pesticides Linked to Bee Declines Might Also Threaten Birds," *Audubon*, Spring 2017, https://www.audubon.org/magazine/spring-2017/the-same-pesticides-linked-bee-declines-might.

63   Christie A. Morrissey, interview with the author, January 20, 2021.

63   Morrissey.

64   E. O. Wilson, "My Wish: Build the Encyclopedia of Life," TED, March 2007, https://www.ted.com/talks/e_o_wilson_on_saving_life_on_earth.

64   Elizabeth Kolbert, "Where Have All the Insects Gone?," *National Geographic*, April 23, 2020, https://www.nationalgeographic.com/magazine/article/where-have-all-the-insects-gone-feature.

67   Margaret Eng, interview with the author, January 2, 2021.

67   Eng.

68   Morrissey, interview.

68   Morrissey.

68   Morrissey.

68–69 "Are the Prairies Getting Quieter? Songbirds Are Declining in Number," CBC News, March 19, 2015, https://www.cbc.ca/news/canada/saskatoon/are-the-prairies-getting-quieter-songbirds-are-declining-in-number-1.3002029.

71   Morrissey, interview.

72   James E. Lovelock, "Hands Up for the Gaia Hypothesis," *Nature* 344, no. 6262 (March 8, 1990): 102.

72   "Scientists Investigate Massive Seabird Die-Off in Alaska," CBS News, January 12, 2016, https://www.cbsnews.com/news/scientists-investigate-massive-seabird-die-off-in-alaska/.

75   Sabrina Shankman, "Dead Birds Washing Up by the Thousands Send a Warning about Climate Change," Inside Climate News, January 15, 2020, https://insideclimatenews.org/news/15012020/seabird-death-ocean-heat-wave-blob-pacific-alaska-common-murre.

75   Shankman.

75   Shankman.

76   John F. Piatt et al., "Extreme Mortality and Reproductive
      Failure of Common Murres Resulting from the Northeast Pacific
      Marine Heatwave of 2014–2016." *PLOS One*, January 15, 2020,
      https://journals.plos.org/plosone/article?id=10.1371/journal.
      pone.0226087.

76   Shankman, "Dead Birds."

77   Piatt et al., "Extreme Mortality."

81   Thomas L. Frölicher, Erich M. Fischer, and Nicolas Gruber,
      "Marine Heatwaves under Global Warming," *Nature* 560, no. 7718
      (August 2018): 364.

81   Jackie Flynn Mogensen, "A New Study about the Death of 1
      Million Seabirds Should Scare the Crap Out of You," *Mother
      Jones*, January 15, 2020, https://www.motherjones.com
      /environment/2020/01/a-new-study-death-of-1-million-common
      -murres/.

82   Douglas W. Tallamy, *Nature's Best Hope* (Portland, OR: Timber,
      2020), 24–25.

87   Douglas W. Tallamy, *Bringing Nature Home: How You Can
      Sustain Wildlife with Native Plants* (Portland, OR: Timber, 2017),
      61.

88   Desirée L. Narango, interview with the author, August 26, 2020.

89   Narango.

89   Narango.

89–90 Narango.

90   Narango.

91   Rene Ebersole, "How to Create a Bird-Friendly Yard," *Audubon*,
      July/August 2013, https://www.audubon.org/magazine/july-august
      -2013/how-create-bird-friendly-yard.

91   Narango, interview.

91–92 Narango.

92   Narango.

93   Tallamy, *Nature's Best Hope*, 62.

94   Sir David Attenborough, *The Life of Birds* (London: BBC Natural
      History, 1998), DVD.

96   "Climate Change and Birds," National Audubon Society,
      September 8, 2014, https://www.youtube.com/watch?v=aN2
      -a82_3mg.

97   A. J. Willingham, "These Black Nature Lovers Are Busting
      Stereotypes, One Cool Bird at a Time," CNN, June 2, 2020, https://
      www.cnn.com/2020/06/03/us/black-birders-week-black-in-stem
      -christian-cooper-scn-trnd/index.html.

100   Morrissey, interview.

Selected Bibliography

A complete bibliography of sources is available at http://rebeccahirsch.com.

Bildstein, Keith L. "A Brief History of Raptor Conservation in North America." In *The State of North America's Birds of Prey*. Edited by K. L. Bildstein, J. Smith, and E. Ruelas. Orwigsburg, PA: Hawk Mountain Sanctuary, 2008.

Eng, Margaret, Bridget J. M. Stutchbury, and Christy A. Morrissey. "A Neonicotinoid Insecticide Reduces Fueling and Delays Migration in Songbirds." *Science* 365 (2019): 1177–1180.

Greenberg, Joel. *A Feathered River across the Sky: The Passenger Pigeon's Flight to Extinction*. New York: Bloomsbury, 2014.

Klem, D., Jr. "Preventing Bird-Window Collisions." *Wilson Journal of Ornithology* 121, no. 2 (2009): 314–321.

Loss, Scott R., Tom Will, and Peter P. Marra. "The Impact of Free-Ranging Domestic Cats on Wildlife of the United States." *Nature Communications* 4, no. 1 (2013): 1396.

Marra, Peter P., and Chris Santella. *Cat Wars: The Devastating Consequences of a Cuddly Killer*. Princeton, NJ: Princeton University Press, 2016.

Narango, Desirée L. "Nonnative Plants Reduce Population Growth of an Insectivorous Bird." *Proceedings of the National Academy of Sciences* 115, no. 45 (2018): 11549–11554.

Narango, Desirée L., Douglas W. Tallamy, and Peter P. Marra. "Native Plants Improve Breeding and Foraging Habitat for an Insectivorous Bird." *Biological Conservation* 213, Part A (2017): 42–50.

Royte, Elizabeth. "The Same Pesticides Linked to Bee Declines Might Also Threaten Birds." *Audubon*, Spring 2017. https://www.audubon.org /magazine/spring-2017/the-same-pesticides-linked-bee-declines-might.

Tallamy, Douglas W. *Bringing Nature Home: How You Can Sustain Wildlife with Native Plants*. Portland, OR: Timber, 2007.

US North American Bird Conservation Initiative. "State of the Birds 2014 Report." Cornell Lab of Ornithology. Accessed August 16, 2021. https:// archive.stateofthebirds.org/state-of-the-birds-2014-report/.

Further Information

Books

Carson, Rachel L. *Silent Spring*. New York: Houghton Mifflin, 1962. Carson offers an eloquent and stark look at the environmental and human health aspects of chemical pesticides, including DDT.

Hirsch, Rebecca E. *Birds vs. Blades? Offshore Wind Power and the Race to Protect Seabirds*. Minneapolis: Millbrook Press, 2017. Follow along as a team of scientists venture into the ocean to track seabirds. Get a close-up look at how researchers are working to ensure that clean, renewable offshore wind power won't spell disaster for the millions of seabirds that play a critical role in ocean food chains.

Hoose, Phillip. *Moonbird: A Year on the Wind with the Great Survivor B95*. New York: Farrar, Straus & Giroux Books for Young Readers, 2012. The author follows B95, a tagged rufa red knot, as the robin-size shorebird bird flies between its wintering grounds in South America and its breeding grounds in the Canadian Arctic. The author explores threats that have reduced the rufa red knot population by nearly 80 percent.

Hughes, Meredith Sayles. *Plants vs. Meats: The Health, History, and Ethics of What We Eat*. Minneapolis: Twenty-First Century Books, 2016. What is the best diet? That's a personal decision, but some modern food-growing practices endanger wildlife. This book explores cultural, ethical, and environmental reasons for choosing certain foods.

Milhaly, Christy, and Sue Heavenrich. *Diet for a Changing Planet: Food for Thought*. Minneapolis: Twenty-First Century Books, 2019. As Earth's human population grows and as climate change disrupts traditional food systems, how can we feed everyone? One solution: eat insects! This book explores insects and other unconventional food sources.

Tallamy, Douglas W. *Nature's Best Hope:*
A New Approach to Conservation
That Starts in Your Yard. Portland,
OR: Timber, 2019. This book is a
call to action, inviting people to take
environmental action into their own
hands by converting their yards into habitat
for birds and wildlife.

Websites

Acopian BirdSavers
https://www.birdsavers.com
Do you want to prevent birds from flying into windows? Zen wind
curtains are a simple, do-it-yourself solution. The curtains break up
reflections in glass so that birds see windows as hard surfaces. Follow
step-by-step directions and watch a video showing how to make them.
Another video, in English and Spanish, shows their use at a bird
observatory in Costa Rica, where the curtains have reduced death
from bird collisions by 99 percent.

Birds and Climate Change Visualizer
https://www.audubon.org/climate/survivalbydegrees
With this Audubon Society resource, you can search your area by zip
code or state and learn what birds are at risk from climate change
where you live.

Live Bird Migration Maps
https://birdcast.info/migration-tools/live-migration-maps
Follow bird migrations in real time during the spring and fall
migration. You can watch actual bird migrations detected by weather
radar. You'll see the migrations begin at sunset across the country
and continue through the night. This site is a partnership among
the Cornell Lab of Ornithology, Colorado State University, and the
University of Massachusetts–Amherst.

Native Plant Finder
https://www.nwf.org/NativePlantFinder/Plants
Discover native plants for bird habitat, ranked by the number of
butterfly and moth species that use them as host plants for their
caterpillars. The information is based on the research of entomologist
Doug Tallamy. You'll get suggestions on what flowers, grasses, trees,
and shrubs will give the most insect food for birds raising their young.

Native Plants Database
https://www.audubon.org/native-plants
Use this database to locate the best native plants to provide habitat for the birds in your area. You'll get suggestions on what trees, shrubs, and flowers to grow, along with what birds each plant may attract.

Seven Simple Actions to Help Birds
https://www.3billionbirds.org/7-simple-actions
Here you can find simple steps you can take to help birds, brought to you by the team of scientists that discovered nearly three billion North American birds had disappeared.

What Birds and Bugs Tell Us about How We Grow Food?
https://www.youtube.com/watch?v=fWV4yHqnZIY
In this TED talk, Christy Morrissey discusses bird studies and what her findings tell us about how agriculture affects birds.

Index

About the Author

Rebecca E. Hirsch has written about science and discovery in dozens of books for children and young adults. Her young adult titles include *Where Have All the Bees Gone? Pollinators in Crisis* and *De-Extinction: The Science of Bringing Lost Species Back to Life*. A former scientist, Hirsch holds a PhD in cellular and molecular biology from the University of Wisconsin. She lives in State College, Pennsylvania, with her family and various pets. You can learn more at her website: www.rebeccahirsch.com.

Photo Acknowledgments

Image credits: Courtesy of National Audubon Society, p. 5; North Wind Picture Archives via AP Images, p. 8; Medvedeva Oxana/Shutterstock.com, p. 11; travelpeter/Shutterstock.com, p. 13; Chrispo/Alamy Stock Photo, p. 18; Richard McMillin/Getty Images, p. 20; Dan Klem via Rachel Hirsch, p. 23; Photo courtesy of USDA Natural Resources Conservation Service, p. 27; TIMOTHY A. CLARY/AFP via Getty Images, p. 30; © Stephen Maciejewski, p. 33; AP Photo/Joseph Kaczmarek, p. 36; SilviaJansen/Getty Images, p. 39; Julija Kumpinovica/Getty Images, p. 43; Hawai'i Department of Land and Natural Resources, p. 44; blickwinkel/Alamy Stock Photo, p. 49; William Leaman/Alamy Stock Photo, p. 50; CBS/Getty Images, p. 56; "The Messenger" SongbirdSOS Productions Inc., p. 59; Federico Rostagno/Shutterstock.com, p. 61; © Margaret Eng via Rebecca Hirsch, pp. 62, 66; Iakov Filimonov/Shutterstock.com, p. 69; AP Photo/Mark Thiessen, p. 73; James Copeland/Shutterstock.com, p. 78; Angel DiBilio/Shutterstock.com, p. 79; DurkTalsma/Getty Images, p. 80; Jacky Parker Photography/Getty Images, p. 83; Raquel Lonas/Getty Images, p. 84; DeeZee/Shutterstock.com, p. 92; JamesBrey/Getty Images, p. 95; AP Photo/Jacqueline Larma, p. 97; AP Photo/Acopian BirdSavers, p. 98; South_agency/Getty Images, p. 101. Design elements: NeMaria/Shutterstock.com; BOONCHUAY PROMJIAM/Shutterstock.com.
Cover and jacket: Norman Bateman/Shutterstock.com.